T0291839

CAMBRIDGE LIBRARY COLLECTION

Books of enduring scholarly value

Botany and Horticulture

Until the nineteenth century, the investigation of natural phenomena, plants and animals was considered either the preserve of elite scholars or a pastime for the leisured upper classes. As increasing academic rigour and systematisation was brought to the study of 'natural history', its subdisciplines were adopted into university curricula, and learned societies (such as the Royal Horticultural Society, founded in 1804) were established to support research in these areas. A related development was strong enthusiasm for exotic garden plants, which resulted in plant collecting expeditions to every corner of the globe, sometimes with tragic consequences. This series includes accounts of some of those expeditions, detailed reference works on the flora of different regions, and practical advice for amateur and professional gardeners.

Commercial Botany of the Nineteenth Century

The nineteenth century witnessed great advances in technology which made transporting natural resources overseas significantly easier. At the centre of a global empire, Britain felt the full economic benefits of introducing and cultivating a range of commercial plants both domestically and in her colonies abroad. First published in 1890, this succinct work by the English botanist John Reader Jackson (1837–1920) surveys these plants. The concise descriptions are enhanced by instructive drawings of significant species. The introduction also contains a chronological table of the century's most important developments in commercial botany. This is followed by chapters organised according to the applications of plants, notably in food, drink, medicine, and the building trade. Jackson points out the species which revolutionised these industries, identifying those at the heart of rapidly growing markets. The coverage includes many commodities which remain commercially significant, such as palm oil, sugar cane, and cotton.

Cambridge University Press has long been a pioneer in the reissuing of out-of-print titles from its own backlist, producing digital reprints of books that are still sought after by scholars and students but could not be reprinted economically using traditional technology. The Cambridge Library Collection extends this activity to a wider range of books which are still of importance to researchers and professionals, either for the source material they contain, or as landmarks in the history of their academic discipline.

Drawing from the world-renowned collections in the Cambridge University Library and other partner libraries, and guided by the advice of experts in each subject area, Cambridge University Press is using state-of-the-art scanning machines in its own Printing House to capture the content of each book selected for inclusion. The files are processed to give a consistently clear, crisp image, and the books finished to the high quality standard for which the Press is recognised around the world. The latest print-on-demand technology ensures that the books will remain available indefinitely, and that orders for single or multiple copies can quickly be supplied.

The Cambridge Library Collection brings back to life books of enduring scholarly value (including out-of-copyright works originally issued by other publishers) across a wide range of disciplines in the humanities and social sciences and in science and technology.

Commercial Botany
of the
Nineteenth Century

JOHN R. JACKSON

CAMBRIDGE
UNIVERSITY PRESS

CAMBRIDGE
UNIVERSITY PRESS

University Printing House, Cambridge, CB2 8BS, United Kingdom

Published in the United States of America by Cambridge University Press, New York

Cambridge University Press is part of the University of Cambridge.

It furthers the University's mission by disseminating knowledge in the pursuit of
education, learning and research at the highest international levels of excellence.

www.cambridge.org
Information on this title: www.cambridge.org/9781108069311

© in this compilation Cambridge University Press 2014

This edition first published 1890
This digitally printed version 2014

ISBN 978-1-108-06931-1 Paperback

Selected botanical reference works available in the
CAMBRIDGE LIBRARY COLLECTION

al-Shirazi, Noureddeen Mohammed Abdullah (compiler), translated by
Francis Gladwin: *Ulfáz Udwiyeh, or the Materia Medica* (1793)
[ISBN 9781108056090]

Arber, Agnes: *Herbals: Their Origin and Evolution* (1938)
[ISBN 9781108016711]

Arber, Agnes: *Monocotyledons* (1925) [ISBN 9781108013208]

Arber, Agnes: *The Gramineae* (1934) [ISBN 9781108017312]

Arber, Agnes: *Water Plants* (1920) [ISBN 9781108017329]

Bower, F.O.: *The Ferns (Filicales)* (3 vols., 1923–8) [ISBN 9781108013192]

Candolle, Augustin Pyramus de, and Sprengel, Kurt: *Elements of the Philosophy
of Plants* (1821) [ISBN 9781108037464]

Cheeseman, Thomas Frederick: *Manual of the New Zealand Flora*
(2 vols., 1906) [ISBN 9781108037525]

Cockayne, Leonard: *The Vegetation of New Zealand* (1928)
[ISBN 9781108032384]

Cunningham, Robert O.: *Notes on the Natural History of the Strait of Magellan
and West Coast of Patagonia* (1871) [ISBN 9781108041850]

Gwynne-Vaughan, Helen: *Fungi* (1922) [ISBN 9781108013215]

Henslow, John Stevens: *A Catalogue of British Plants Arranged According to
the Natural System* (1829) [ISBN 9781108061728]

Henslow, John Stevens: *A Dictionary of Botanical Terms* (1856)
[ISBN 9781108001311]

Henslow, John Stevens: *Flora of Suffolk* (1860) [ISBN 9781108055673]

Henslow, John Stevens: *The Principles of Descriptive and Physiological Botany*
(1835) [ISBN 9781108001861]

Hogg, Robert: *The British Pomology* (1851) [ISBN 9781108039444]

Hooker, Joseph Dalton, and Thomson, Thomas: *Flora Indica* (1855)
[ISBN 9781108037495]

Hooker, Joseph Dalton: *Handbook of the New Zealand Flora* (2 vols., 1864–7)
[ISBN 9781108030410]

Hooker, William Jackson: *Icones Plantarum* (10 vols., 1837–54)
[ISBN 9781108039314]

Hooker, William Jackson: *Kew Gardens* (1858) [ISBN 9781108065450]

Jussieu, Adrien de, edited by J.H. Wilson: *The Elements of Botany* (1849)
[ISBN 9781108037310]

Lindley, John: *Flora Medica* (1838) [ISBN 9781108038454]

Müller, Ferdinand von, edited by William Woolls: *Plants of New South Wales*
(1885) [ISBN 9781108021050]

Oliver, Daniel: *First Book of Indian Botany* (1869) [ISBN 9781108055628]

Pearson, H.H.W., edited by A.C. Seward: *Gnetales* (1929)
[ISBN 9781108013987]

Perring, Franklyn Hugh et al.: *A Flora of Cambridgeshire* (1964)
[ISBN 9781108002400]

Sachs, Julius, edited and translated by Alfred Bennett, assisted by W.T. Thiselton
Dyer: *A Text-Book of Botany* (1875) [ISBN 9781108038324]

Seward, A.C.: *Fossil Plants* (4 vols., 1898–1919) [ISBN 9781108015998]

Tansley, A.G.: *Types of British Vegetation* (1911) [ISBN 9781108045063]

Traill, Catherine Parr Strickland, illustrated by Agnes FitzGibbon Chamberlin:
Studies of Plant Life in Canada (1885) [ISBN 9781108033756]

Tristram, Henry Baker: *The Fauna and Flora of Palestine* (1884)
[ISBN 9781108042048]

Vogel, Theodore, edited by William Jackson Hooker: *Niger Flora* (1849)
[ISBN 9781108030380]

West, G.S.: *Algae* (1916) [ISBN 9781108013222]

Woods, Joseph: *The Tourist's Flora* (1850) [ISBN 9781108062466]

For a complete list of titles in the Cambridge Library Collection please visit:
www.cambridge.org/features/CambridgeLibraryCollection/books.htm

COMMERCIAL BOTANY

OF THE

NINETEENTH CENTURY.

COMMERCIAL BOTANY

OF THE

NINETEENTH CENTURY.

A Record of Progress in the Utilisation of Vegetable Products
in the United Kingdom, and the Introduction of
Economic Plants into the British Colonies,
during the Present Century.

BY

JOHN R. JACKSON, A.L.S.,

Associate Royal Botanical Society of London, Curator of the
Museums, Royal Gardens, Kew.

———————•♦•———————

CASSELL & COMPANY, LIMITED:

LONDON, PARIS & MELBOURNE.

1890.

CONTENTS.

COMMERCIAL BOTANY OF THE NINETEENTH CENTURY.

INTRODUCTION.

In considering this subject, the whole history of Economic Botany may be said to be placed under review, for it is quite within the last thirty years that anything like real or general attention has been directed to the subject.

It is true that in the present century no single plant has been introduced either to commerce or for home cultivation of such widespread importance as the tobacco and potato plants, nevertheless what has been accomplished in a comparatively few years in the cultivation of the cinchonas and the various caoutchouc-producing plants in various parts of the world will bear favourable comparison with anything done in a similar direction in previous centuries, and judging from the present rate of scientific progress the importance of these plants alone in future years may, and probably will, equal those of the tobacco and potato.

It would be impossible to form any correct idea of what has been attained in the knowledge of plants, useful or otherwise, without referring to the results of the principal expeditions which have left our shores for different parts of the world during the present century, such, for instance, as Ross's Antarctic Expedition, which . resulted in "The Botany of the Antarctic Voyage of H.M. Ships *Erebus* and *Terror*, in the years 1839 to 1843," by Dr. (now Sir)

B

J. D. Hooker ; or Captain Kellet's voyage of the *Herald*,
after which appeared "The Botany of H.M.S. *Herald*
during the years 1845 to 1851," by Berthold Seemann ; or
in still later times Captain Nares' *Challenger* Expedition
from 1873 to 1876, the botany of which occupies two large
volumes, principally the work of Mr. W. B. Hemsley,
F.R.S. Not that these expeditions have resulted directly in
the introduction of any one useful plant either for general
culture or commerce, but they have been instrumental in im-
parting a knowledge of the resources of the several countries
visited, and in this way have awakened an interest in them.
Important, indeed, as these expeditions have been in elu-
cidating the botany of the world, still more so has been the
formation of the several museums in the principal centres of
the United Kingdom for the especial purpose of developing
the economic resources of the vegetable, animal, and mineral
kingdoms, such as the Food Collection, first at South
Kensington in 1857 and later at Bethnal Green, the Industrial Museum at Edinburgh, and the Museums of
Economic Botany at Kew, founded in 1847. These, together with the Royal Botanic Society of London, founded
in 1839, and the Pharmaceutical Society of Great Britain,
founded in 1841, must always be considered the centres
from which knowledge on these points has flowed, and
continues to flow. Nor must we forget the several International Exhibitions since 1851, where the vegetable
resources of the globe, especially of our colonies, have been
prominently brought to the notice of millions of people.
Then, in connection with these museums and exhibitions
is the literature which emanates from them, such as the
handbooks and guides, in which, though published mostly
for a few pence, a mass of valuable information is given.
We cannot leave this part of the subject without a word of
high commendation on the handbooks and catalogues issued
by the several colonies at the Colonial and Indian Exhibition

of 1886, which should be in the library of everyone interested in Economic Botany.

In the following notes the difficulty has been to keep them within what might be considered reasonable bounds. It was found impossible to even enumerate all the plants reputed to have economic properties that have been introduced to the United Kingdom during the present century, therefore those that seemed to have but little claim for notice have been omitted. On the other hand, such important plants as those yielding india-rubbers, gutta-perchas, cinchonas, etc., whose cultivation in other countries than their own is of vast importance to our commerce, and to the prosperity and welfare of our countrymen in our widely-spread dominions, have received a large share of attention, because it was thought that the information here brought together would be useful were it more widely disseminated than it has hitherto been.

Throughout these pages it will be seen how often fresh products have been brought forward and have shown promise of becoming important commercial articles, and then have collapsed, sometimes finally, and sometimes to crop up again after a lapse of years. It is hoped that a perusal of these facts will inspire those who have opportunities to take up new products, or even to resuscitate those recorded here as having failed, to persevere in thoroughly testing their properties, or placing them in the proper channels for so doing.

A commercial rather than a scientific arrangement has been adopted, as being probably the more generally useful.

CHRONOLOGICAL TABLE OF PLANTS.

1801.—Waste vegetable fibres applied to paper-making.
1806 —Rhatany root (*Krameria triandra*) introduced as a medicine.
1807.—Rohun bark (*Soymida febrifuga*) introduced as a medicine.
Gambier or Terra japonica (*Uncaria Gambier*) introduced about this time.

B 2

4 COMMERCIAL BOTANY.

1809.—Quassia wood (*Picræna excelsa*) introduced as a medicine.
Belladonna leaves (*Atropa Belladonna*) introduced as a medicine.
Cowhage (*Mucuna pruriens*) introduced as a medicine.
1813.—Cajuput oil (*Melaleuca leucadendron*, var. *minor*) introduced about this time.
Patent granted for an invention for making fabrics air-proof by being treated with India-rubber or Caoutchouc.
The use of Ipecacuanha as a specific in dysentery confirmed in England.
1819.—The development of the India-rubber trade commenced about this time.
1820.—Colchicum seeds (*Colchicum autumnale*) introduced as a medicine.
1821.—Buchu leaves (*Barosma crenulata, B. serratifolia,* and *B. betulina*) introduced as a medicine.
1825.—Rusa or ginger grass oil (*Andropogon Schœnanthus*) introduced about this time for perfumery.
1826.—Vegetable ivory (*Phytélephas macrocarpa*) introduced about this period.
1829.—Indian tobacco (*Lobelia inflata*) introduced as a medicine.
1832.—Lemon grass oil (*Andropogon citratus*), introduced for perfumery.
1835.—Greenheart bark (*Nectandra Rodiœi*) first received attention as a tonic and febrifuge.
1837.—Beetroot Sugar Refinery established at Chelsea.
Gama grass or buffalo grass (*Tripsacum dactyloides*) introduced for fodder.
1839.—Cherry-laurel (*Prunus Laurocerasus*) introduced for making cherry-laurel water.
Cultivation of cinchona plants suggested in India.
From this period to 1841 Kousso (*Hagenia abyssinica*) attracted some attention as an anthelmintic.
1840.—Tibet hay (*Prangos pabularia*) introduced as a fodder plant.
Ordeal bean of Old Calabar (*Physostigma venenosum*) introduced.
Ground nut (*Arachis hypogœa*) introduced.
First vineyard planted in S. Australia.
1842.—Tussock grass (*Aira flabellata = Dactylis cæspitosa*) introduced as a fodder plant.
Jute (*Corchorus capsularis*) introduced about this time.
Gutta-percha first brought to notice.

1844.—Introduction of glycerine to commerce.
1847.—Cotton seeds first imported as oil seeds, and piassaba fibre introduced about this time for brush-making.
1849.—" Child's night-lights " introduced.
Masseranduba (*Mimusops elata*) milk and China grass (*Bœhmeria nivea*) first brought to notice about this time.
1850.—Cedron (*Simaba Cedron*) introduced as a medicine.
Sumbul (*Ferula* [*Euryangium*] *Sumbul*) introduced as a medicine.
Bael fruit (*Ægle Marmelos*) introduced as a medicine.
1851.—Experiments made in cultivation of *Ullucus tuberosus* as a substitute for the potato.
Shea butter (*Butyrospermum Parl.*) introduced about this time for soap-making.
1852.—First lime-fruit orchards planted in Montserrat.
1854.—Mexican fibre (*Agave heteracantha*) introduced about this time for brush-making.
1856.—Paper first made from Esparto.
Blue gum (*Eucalyptus Globulus*) introduced for cultivation, and for its medicinal properties.
Guarana (*Paullinia sorbilis*) introduced, and again in 1870.
1858.—Larch bark (*Larix europea*) first used as a medicine.
1859.—Balata (*Mimusops globosa*) introduced.
Udika fat (*Irvingia Barteri*) first brought to notice.
Expedition arranged under Mr. Clements Markham to proceed to the South American Forests to collect cinchonas for transmission to India.
1860.—*Urceola esculenta* first noticed as a rubber-yielding plant.
Black snake root (*Cimicifuga racemosa*) introduced as a medicine.
Belladonna root (*Atropa Belladonna*) introduced as a medicine.
Pine wool introduced to commerce about this time.
1861.—Briar-root pipes introduced about this time.
1862.—Palmite (*Prionium palmita*) introduced for brush and paper-making.
Indian poke root (*Veratrum viride*) introduced as a medicine.
1863.—Wild black cherry bark (*Prunus serotina*) introduced as a medicine.
1864.—*Podophyllum peltatum, Mallotus philippinensis,* and *Hemidesmus indicus* admitted to British Pharmacopœia.
1866.—Elands Bontjes (*Elephantorrhiza Burchellii*) first brought to notice, but not used in this country for tanning till 1886.

1867.—About this time attention began to be directed by the Indian Government to the introduction and cultivation of Ipecacuanha in India.

Roffia (*Raphia Ruffia*) introduced for tying plants about this time.

Betel nuts (*Areca Catechu*) first used in medicine.

1868.—Botanical source of Zanzibar Anime determined.

1869.—Sumbul plant (*Ferula Sumbul*) discovered in Samarkand.

Coffee plantations attacked in Ceylon by *Hemileia vastatrix*.

1870.—Jalap (*Ipomœa purga*) first cultivated in India.

Strophanthus first brought to notice.

Telfairea occidentalis seeds first received at Liverpool as oil seeds.

Ispaghul (*Plantago ovata* or *Ispaghula*) introduced.

Quantities of red cinchona bark from Darjiling arrive in the London market.

1871.—Cundurango (*Marsdenia cundurango*) introduced for the cure of cancer.

1873.—Para rubber plants (*Hevea brasiliensis*) introduced to India.

Pituri (*Duboisia Hopwoodi*) introduced to medicine.

Rheum officinale introduced to cultivation in this country.

Mahogany (*Swietenia Mahagoni*) introduced for cultivation in India.

Experiments made at Chatham with a paint prepared with Euphorbia juice.

1874.—*Cinnamodendron corticosum* flowered for the first time in England at Royal Botanic Gardens, Regent's Park.

Coca (*Erythroxylon Coca*) first brought prominently to notice in this country.

Goa powder (*Andira Araroba*), Jaborandi (*Pilocarpus pennatifolius*), Boldo (*Peumus Boldus*), and Damiana (*Turnera diffusa*) introduced to medicine.

Liberian coffee (*Coffea liberica*) introduced for cultivation in Jamaica, Bahamas, Barbados, Bermuda, Dominica, Montserrat, New Granada, Rio de Janeiro, Mauritius, Natal, Ceylon, East Indies, and Java.

1875.—Paper first made from Bamboo.

Algaroba (*Cæsalpinia brevifolia*) introduced for tanning.

Sumbul plant flowered at Moscow, and scientific name determined.

Seeds of Central American rubber plants (*Castilloa*) collected for transmission to India.

Carnauba roots (*Copernicia cerifera*), Caroba leaves (*Cybistax antisyphilitica*), and Dita bark (*Alstonia scholaris*) introduced to medicine.

1876.—Paper first made from Baobab bark.

Uniola virgata tested for paper-making.

Bamia cotton introduced.

Kava (*Piper methysticum*) began to attract some attention for its medicinal properties.

Seeds of *Cassia occidentalis* introduced as a substitute for coffee.

Ipecacuanha dried and prepared for use in India from native-grown plants.

Central American rubber plants (*Castilloa elastica*) introduced into West Africa, Ceylon, Java, etc.

Para-rubber plants (*Hevea brasiliensis*) introduced into West Africa, Dominica, Jamaica, Java, Queensland, Singapore, and Trinidad.

Ceara-rubber plants (*Manihot Glaziovii*) introduced to Kew for transmission abroad.

1877.—First crop of cinchona bark received in London from Jamaica plantations.

Attention first drawn to Mahwa flowers (*Bassia latifolia*) for feeding cattle, and for distilling purposes.

Attempts first made to introduce African rubber plants for transmission abroad.

Prickly comfrey (*Symphytum peregrinum*) introduced as a fodder plant.

Calotropis gigantea, Cavanillesia platanifolia, and *Yucca brevifolia* suggested as paper materials.

1878.—*Molinea ca, lea* and *Ischæmum angustifolium* introduced for paper-making.

Pods of *Wagatea spicata* introduced from India for tanning.

"Zulu" hats from *Cyperus tegetiformis* introduced.

Queensland Fever Bark (*Alstonia constricta*) and Chaulmugra (*Gynocardia odorata*) introduced to medicine.

Wourali poison (*Strychnos toxifera*) first brought to notice in this country.

Liberian rubber plants (*Ficus Vogelii*) introduced.

About this time a considerable amount of attention was given to the properties of the Papaw (*Carica Papaya*).

1879.—White Quebracho (*Aspidosperma Quebracho-blanco*), Yerba Reuma (*Frankenia grandifolia*), Menthol, and Tonga introduced to medicine.

Paper Mulberry bark (*Broussonetia papyrifera*) and Rye straw first used for paper-making.

Chestnut flour (*Castanea sativa*) brought to notice as a probable article of food.

Sugar made in America from *Sorghum saccharatum*.

African rubber plants (*Landolphia*) introduced into Australia, Ceylon, North America, Demerara, Fiji, Jamaica, Rio de Janeiro, Natal, Singapore, and Trinidad.

1880.—Cocoa plants (*Theobroma Cacao*) transmitted from Trinidad for acclimatisation in Ceylon, Singapore, and Fiji.

Socotra Aloes plant introduced to cultivation.

Ledger bark (*Cinchona Calisaya*, var. *Ledgeriana*) attracts considerable attention.

China Cuprea (*Remijia spp.*) appears in the English market.

Artificial Indigo introduced.

First consignment of Indian boxwood received.

Lallemantia iberica seeds introduced as oil seeds.

Gutta Sundek seeds introduced from Perak for cultivation in Ceylon.

Kola-nut plant (*Cola acuminata*) propagated at Kew and transmitted to India, Ceylon, Java, Singapore, Demerara, Dominica, Mauritius, Sydney, and Zanzibar.

1881 —Anda-assu (*Jcannesia princeps*) and Cheken (*Myrtus Cheken*) introduced to medicine.

Jalap (*Ipomœa purga*) successfully cultivated in Jamaica.

Plantain and Banana stems proposed for paper-making.

Myristica surinamensis seeds imported as oil seeds.

Jequirity (*Abrus precatorius*) and *Euphorbia pilulifera* introduced to medicine.

1883.—Manaca (*Franciscea uniflora*) and Cascara Sagrada (*Rhamnus Purshiana*) introduced to medicine.

Ogea Gum (*Daniellia sp.*) introduced.

Discovery of Inhambane Copal (*Copaifera Gorskiana*).

Kittool fibre (*Caryota urens*) first used commercially for brush-making.

Seeds of *Hyptis spicigera* imported as oil seeds.

About this time Paraguay tea (*Ilex paraguariensis*) was introduced as a substitute for Chinese tea.

1884.—Chew stick (*Gouania domingensis*) introduced to medicine.

Seeds of *Myristica angolensis* and *Polygala rarifolia* imported as oil seeds.

1885.—Cape boxwood (*Buxus Macowani*) introduced.

Doundake bark (*Sarcocephalus esculentus*), Mabee bark (*Colubrina reclinata*), and Cascara Amarga (*Picramnia antidesma*) introduced to medicine.

Choco (*Sechium edule*) introduced into Ceylon, India, and Singapore.

1886.—Fresh fruits of various kinds from West Indies, British Guiana, Australia, and North America received at Colonial and Indian Exhibition, and sold in the Colonial market.

Piche (*Fabiana imbricata*) introduced to medicine.

About this time Gum Euphorbium was introduced for mixing with rubber.

1887.—*Lophira alata* seeds introduced as oil seeds.

Shantung cabbage (*Brassica chinensis*) grown at Kew.

Crosnes tubers (*Stachys tuberifera*) grown in this country and introduced as a vegetable.

1888.—Demerara rubber (*Forsteronia gracilis*) and Jamaica rubber (*F. floribunda*) introduced for trade.

Brazilian gum arabic (*Piptadenia macrocarpa*) introduced to commerce.

Bombay aloe fibre brought to notice at Kew.

Jarrah wood (*Eucalyptus marginata*) first used for road-paving in London.

1889.—Bolobolo fibre (*Honckenya ficiforia*) introduced from Lagos.

Madagascar and Lagos Piassaba first brought to the notice of the Kew authorities, though the former appears to have been known in commerce for some years.

CHAPTER I.

INDIA-RUBBER OR CAOUTCHOUC.

FEW, if any, products, vegetable or otherwise, have made such rapid advances in any given time as has caoutchouc or India-rubber. When we remember that it was quite unknown in this country till the latter end of the last century, and when we try to think of what the world would now be without it, we are able to realise to some extent the great value of this remarkable and interesting substance.

The introduction of Para-rubber antedates by some years the period within our review. It will perhaps be of some use to briefly sketch the history of the development of caoutchouc in this country.

In a work on "Perspective," by Dr. Priestley, published about 1770, the writer speaks of the rubber as a new and important discovery for "wiping from paper the marks of a black-lead pencil," and he says that it could then be obtained at only one place in London, the price being three shillings for a cubical piece of about half an inch. In 1836–37 Para-rubber of good quality was imported into this country to the extent of 141,735 pounds, which had increased twenty years later to 3,477,445 pounds.

The first patent granted in the present century in connection with caoutchouc was in 1813 to John Clark, for an invention by which fabrics treated with India-rubber were made air-tight and applicable for air-beds, pillows, cushions, etc. ; but for the greater improvements in the India-rubber manufactures we are indebted to Mr. Thomas Hancock, so long connected with the firm of Charles Macintosh and Co. ; and though the commencement of the trade in this country dates from about the year 1819, its greater development has been effected within the last thirty years. In 1839 India-rubber was first applied to the manufacture of large flat-bottomed boats, which were used for pontoons as well as for exploring purposes, their great recommendation being their portable character and the ease with which they could be carried when not inflated.

The patents that have been granted for manufactures in connection with India-rubber have been extremely numerous. Some idea of their extent may be obtained when it is stated that the "Abridgments of Specifications relating to India-rubber (Caoutchouc and Gutta-percha)" issued by the Patent Office forms an octavo volume of over 700 pages, and this extends only to the end of the year. 1857. The use of rubber is now so universal that it has come to be regarded as one of the necessaries of life. Enveloped in a thin film of India-rubber, as we are when equipped in our macintosh, we are inclined to regard it as one of the blessings of modern times. Air-cushions, hot-water bottles, water-beds, elastic stockings, door-mats, etc., are all modern luxuries, the introduction of which we owe to caoutchouc. Amongst other uses to which soft rubber is put may be mentioned valves, buffers, washers, packing, garden hose, waterproof garments, etc.; and from hard rubber or ebonite are manufactured photographic and surgical instruments, vessels for holding liquids, acid-pumps, insulators, and a host of other articles both useful and ornamental. Vulcanite or hard rubber has indeed almost entirely replaced glass for frictional electrical machines.

Important as all these applications are, and to meet the demands for which enormous quantities of rubber are now brought into the English market, they are eclipsed by the application of this substance as an insulator in telegraphy. The progress made even in the last ten years in the development of electrical science has been materially assisted by the peculiar properties of India-rubber, for as a coating for the wires of deep-sea telegraphs and telephones it, together with its allied substance gutta-percha, has played a most conspicuous part. As an insulator, however, India-rubber possesses a superiority over gutta-percha in its much higher resistance and considerably lower capacity. For torpedoes and military telegraphs, again, it is largely used, because it is not so readily injured as gutta-percha by rough handling or by heat. One of the most recent adaptations of rubber is in the manufacture of kamptulicon for floor-coverings, which is composed of waste rubber and cork dust.

PARA-RUBBER (*Hevea brasiliensis*).—In the earlier years of the rubber industry the supply came almost exclusively from Para. Other sources have, however, since been discovered, and at the present time the rubber supplies to this country are procured from various plants of South America, Africa, India, Borneo, and the Malay Archipelago. In consequence of the very great demand for this substance, and the fear lest the sources of supply should become exhausted, the attention of the Kew authorities was first drawn in 1873 to the necessity of introducing the Para-rubber plant to India. In the "Report on the Progress and Condition of the Royal Gardens at Kew" for that year the following paragraph occurs :—" Dr. King, the Superintendent of the Calcutta Botanic Gardens, has returned to his duties, taking with him living plants of the true India-rubber plant of Para (*Hevea brasiliensis*), the seeds of which were procured from the Amazons and sent to Kew by Mr. Markham of the India Office." Again, in the report for the following year (1874), Dr (now Sir Joseph) Hooker writes :—" The plants of the true India-rubber of Para (*Hevea brasiliensis*), which I stated in my last report had been taken out to India by Dr. King, Superintendent of the Calcutta Botanic Gardens, have safely arrived, and have already to some extent been propagated by cuttings. The propagation of this tree is extremely important, not merely from the valuable quality of the rubber obtained from it, but also in view of the diminished supply from the Indian *Ficus elastica*, which, owing to its epiphytal germination and mode of growth, is not well adapted for cultivation for this purpose, while severe inroads have been made upon it in the forests where it occurs."

Again, in 1875 Sir Joseph Hooker reported that experiments at Kew proved that the *Hevea brasiliensis* was capable of easy propagation by cuttings, and that the seeds very quickly lost their germinating power, hardly one per cent. reaching Kew alive.

In 1876 Sir Joseph Hooker reports that, " On the 14th of June Mr. H. A. Wickham, a resident on the Amazons, who has been commissioned by the India Office to collect seeds of *Hevea brasiliensis*, arrived in England with 70,000, obtained on the Rio Tapajos. In consequence of their retaining vitality for but a very short period, they were all sown the day after arrival, and covered a space when sown of over 300 square feet, closely packed together. About $3\frac{3}{4}$ per cent. germinated—some as early as the fourth day after sowing, and many in a few days reached a height of 18 inches. Upwards of 1,900 plants were transmitted on August 12th, in .38 Wardian cases made specially to accommodate the rapid growth of the seedlings, to Ceylon under the charge of a gardener." Of this consignment 90 per cent. of the plants reached Ceylon in excellent condition, where they were nursed and established for subsequent transmission through the Indian Gardens to Assam, Burma, and other provinces.

Smaller consignments were also made to the West Coast of Africa, Dominica, Jamaica, Java, Queensland, Singapore, and Trinidad. In this year Mr. Robert Cross, who had been sent to South America to collect the living plants, arrived with about 1,000, a very small proportion of which ultimately survived.

In the following year an unsuccessful attempt was made to transmit some 50 of the plants direct to Burma; all of

them, however, perished on their journey. Plants were
afterwards sent to Saharanpur, Calcutta, Assam, and to
Burma. Subsequent reports from Ceylon showed that the
climate was apparently suited to the growth of the species,
forming handsome spreading trees, from which cuttings were
taken in large quantities, and after being struck were distributed. Plants were sent from Singapore to Perak, and
Mr. Low, reporting upon them under date February 3rd,
1879, says that "the *Heveas* are now 12 to 14 feet high.
They take to the country immensely." From Burma the
reports down to 1879 were favourable, but from Fiji, Calcutta, Assam, Zanzibar, and Jamaica they have not been
assuring. It will be seen from these facts that no pains
have been spared to introduce into the British colonies and
dependencies a really useful plant for extended cultivation;
and though the result has not been so satisfactory as was
anticipated, the introduction and cultivation of other plants
and the discovery of other sources of supply of rubber has
emanated from these efforts. But notwithstanding these
discoveries Para-rubber is still the principal source of supply,
the total imports of which were estimated during the year
1889 to amount to nearly 6,000 tons.

CENTRAL AMERICAN RUBBER (*Castilloa elastica*).—Next
to Para-rubber the plants that have received most attention
for the purpose of cultivation in other countries are those
which yield the kinds known as CENTRAL AMERICAN RUBBER
(*Castilloa elastica*) and CEARA SCRAP (*Manihot Glaziovii*).
The first-named began to attract the attention of the authorities of Kew in 1875. In the "Kew Report" for that year
it is stated that "Mr. Robert Cross was despatched by the
India Office to Central America to obtain seeds and plants"
of this species. Upwards of 7,000 seeds were transmitted
to Kew, but all failed to germinate. With a good deal of
difficulty, and after undergoing shipwreck, Mr. Cross
succeeded in bringing home a considerable collection of

cuttings, which were propagated so rapidly that in 1876 they were ready for distribution to the West Coast of Africa, Ceylon, and Java. From the two latter places satisfactory reports have been received. Dr. Trimen, reporting from Ceylon in 1880, says :—" Our largest trees at Heneratgoda have now a circumference of nearly 17 inches at a yard from the ground, and the trees are beginning to take their true form." Again, in October, 1882, Dr. Trimen writes :—" We have some sturdy little seedlings of *Castilloa* coming on from seed. Only three fruits ripened in June, and the fifteen seeds from these were sown at once, and germinated in fifteen days. The rubber from *Castilloa* strikes me as the most satisfactory sort growing here, the proportion of caoutchouc in the milk being larger than in any of the others."

Of CEARA-rubber, plants and seeds were brought home by Mr. Cross to the Royal Gardens, Kew, in 1876. The botanical name of this valuable rubber-plant was unknown up to this period.

Under the name of *Hevea guyanensis*, a plant had been in cultivation in the gardens of the Royal Botanic Society, Regent's Park, as well as at Buitenzorg, Java, and at Mauritius, which was proved in 1877 at Kew to be *Manihot Glaziovii*, the same species brought home by Mr. Cross as the Ceara-rubber plant. These plants were rapidly propagated at Kew, and quantities were sent to Singapore, Calcutta, and Ceylon, and in 1878 to the Conservator of Forests in Madras. From this period the distribution of Ceara-rubber plants was proceeded with rapidly from most of the Colonial Botanic Gardens, which had received their first consignments from Kew. So important and successful has the spread of this species become that the insertion here of a few extracts on the subject, from the reports of the officers in charge of the several gardens to which they were originally sent, may be desirable.

In 1879 Dr. King, of Calcutta, reported as follows:—
"The Ceara-rubber promises to grow well in Calcutta. The seedlings received from Kew have thriven vigorously, and some of them are now 20 feet high. The Director of the Botanic Garden in Ceylon having, at the request of the Secretary of State, undertaken the propagation of this species, a quantity of seeds of it were distributed by him to Indian officers during the year. Supplies were, I understand, sent to the Conservators of Forests in Burma and Assam, and to the Inspector-General of Forests for Madras. A large supply was received at this garden, and a thousand seeds were sent, at the request of the Conservator of Forests of Bengal, to the officers in charge of the forest plantations near Chittagong. The seeds received here have begun to germinate, and I expect before long to be in a position to issue supplies of seedlings for trial in different parts of the country. The plant appears to thrive very well in Upper India; and if the quality of the rubber yielded by it in this country turns out to be good, its introduction may prove to be of much importance." In 1880 as many as 24,550 seeds and 1,879 rooted cuttings of this important plant had been sent out to different parts of the world from the Botanic Garden, Ceylon. In 1882 Dr. Trimen reported from Ceylon that some planters were going in largely for cultivating this plant, and that if it proved profitable it would be "a great help, as it grows anywhere up to almost 2,500 feet." From a quotation from the *Madras Mail* of October 24, 1883, published in the "Kew Report" issued in November, 1883, some idea of the progress of the plant in Southern India may be gathered. "About six months ago," the writer says, "some gentlemen imported Ceara-rubber seed from Ceylon. The produce of these trees may now be seen flourishing in a wonderful manner at the foot of the Nilgiri Hills. . . . The rapid growth of the Ceara-rubber tree is marvellous; some, measured six months old

from seed, were fully eight feet high, and a cutting that had been put down scarcely six months ago was quite eight feet high and in blossom. Being of such wonderfully rapid growth, the tree is naturally very susceptible to wind and liable to be blown over, until it gets firm hold of the ground. Consequently a sheltered position is most neces-

INDIA-RUBBER (*Hevea brasiliensis*).

sary. It seems to thrive on poor soil, requires no shade and very little rain. . . It is better to plant out the young plants after the first heavy burst of the Monsoon—say in the months of August or October—when the ground is thoroughly saturated, and the showers only occasional, with bursts of sunshine between. The germination of the seed seems a very simple process, and generally occurs in ten days, and sometimes less, from the time the seed is placed in the damp sand. The seed-coat, being extremely hard,

C

requires very careful filing, so as to enable it to burst more easily; if this is not done, the seed may take months to germinate."

A new source of India-rubber has quite recently been brought to the notice of the Kew authorities under the name of COLOMBIAN RUBBER, or COLOMBIA VIRGEN. By this latter name it seems to have been known in English commerce for the last few years, though the origin or source of it has only just been determined at Kew as that of *Sapium biglandulosum*, a euphorbiaceous plant allied to *Manihot Glaziovii*, yielding Ceara scrap (*see* pp. 14 and 15). The details connected with the discovery of this rubber are fully given in the *Kew Bulletin* for July, 1890. It is there said that Colombian Rubber has been generally known in commerce from the place of export as " Carthagena," and is supposed to have been the produce of species of *Castilloa*, which may to some extent still be the case. The plants yielding this so-called "Colombia Virgen " rubber, unlike all other known sources of this substance, grow at a high elevation on the Colombian Andes, viz., at from 6,000 to 8,000 feet above the sea. A report of an examination of this rubber made at the works of the India-rubber, Gutta-percha, and Telegraph Works Company, Limited, at Silvertown, was not favourable, in consequence of the presence of a resinous substance; but a more recent opinion of a well-known firm of rubber brokers of London and Liverpool, is that it " is of a very superior quality indeed," and was valued in May, 1890, at about 2s. 11d. or 3s. per pound.

AFRICAN RUBBER (species of *Landolphia*).—A large quantity of rubber has found its way into the English market for a long time under the name of African rubber, exported chiefly from the West Coast. It was known to be produced from some species of *Landolphia*, climbing plants with thick, woody stems, belonging to the natural

order Apocynaceæ ; the species, however, which yield the
rubber were but imperfectly understood, nevertheless attempts were made in 1877 to introduce the rubber-producing
species to Kew.   Sir John Kirk, H.M. Consul-General at
Zanzibar, thus writes in the "Kew Report" for 1877 :
"The district called Mungao extends from lat. 9° 25' to
Delgado, in lat. 10° 41.   This last year yielded £90,000
worth of India-rubber, an industry that has been created in
the last two years by my representation.   This year the
yield will be more, and other places are now collecting it.
Thus, Kilwa and Mombasa will this year probably double
the supply, which I anticipate will reach in value not less
than £180,000 worth of India-rubber.   East Africa to the
south—that is, from Delgado Bay to the Zambesi—is producing it as well."   Two years later, in 1879, a considerable stock of these plants were got together at Kew, and
were distributed to the Botanic Gardens of Adelaide, Brisbane, Cambridge, U.S., Ceylon, Demerara, Fiji, Jamaica,
Natal, Rio de Janeiro, Singapore, Sydney, Toronto, and
Trinidad.   From the material thus collected at Kew, the
authorities were enabled to clear up the doubts regarding
the identity of the species, both from the East and West
Coasts, which yield rubber, the results of which were published in the "Kew Report" for 1880.   These notes are a
very important contribution to the knowledge of African
rubber-yielding plants, and to the sources of the commercial
supply of this commodity, as they proved that from the
genus *Landolphia* the whole of the East and West African
rubber is obtained.   The following species may be enumerated :—

L. owariensis.—This species possesses the widest lati-
tudinal range, having been collected in Sierra Leone,
Angola, Niger, near the mouth of the Congo, and under a
slightly different form it was found in North Central Africa
by Schweinfurth, who remarks that " it is well known in
    c 2

the Guinea trade for its production of caoutchouc." He
further describes the fruit as having a sourness exceeding
that of the citron, in consequence of which the natives
of Djur-land prepare from it a "beverage refreshing as
lemonade."

L. florida.—This species is widely distributed over the
whole of Central Tropical Africa, and yields a portion of
the rubber obtained both from the East and West Coasts.
irrespective of its commercial interest, Sir Joseph Hooker
says: "Its heads of large, sweet-scented, jasmine-like
flowers would render it a desirable introduction, from
merely a horticultural point of view, in stove cultivation."
The plant flowered at Kew in 1887. The fruits, which
are about the size and shape of a pomegranate, are very
acid, and are eaten by the natives on the Niger. West
African rubber, which comes into the English market in
masses composed of more or less agglutinated small cubes,
is collected in the following peculiar manner:—When
wounded, every part of the plant exudes a milky juice,
which, however, does not flow, but dries quickly, so as to
form a ridge over the wound and to prevent the further
flow. Long cuts are made in the bark by the native
collectors, and as the milky juice flows out it is continually
wiped off with the fingers and smeared over the arms,
shoulders, and breast till a thick covering is formed, when
it is peeled off their bodies and cut into small squares,
which are said to be afterwards boiled in water.

Regarding the produce of rubber from this species on
the East Coast of Africa, Sir John Kirk, reporting to the
Foreign Office from Zanzibar in 1879, says : "It has been
for many years known that the jungles on the mainland
contained almost anywhere an abundance of rubber-producing lianas, and on my first arrival at Zanzibar an
attempt was made, at my representation, to induce the
people of Dar Salam to collect it; at that time, however,

everyone was engaged in the slave trade, and the experiment in consequence failed. Within the last few years India-rubber has, however, become one of the chief exports we possess."

L. Kirkii.—This species, which is named in honour of Sir John Kirk, who discovered it and sent it to Kew, is the source of the best quality and the largest quantity of rubber obtained on the Zanzibar coast. The development of this trade is due entirely to the energy of Sir John Kirk. Some idea of its progress may be obtained from the following extracts from reports made to the Foreign Office in 1880 by Vice-Consul Holmwood and Consul O'Neill. The former describes the mode of collecting as follows : "The process consisted in cutting clean slices of bark from the trunk and branches from 3 to 10 inches in length, and from $\frac{1}{2}$ to $\frac{3}{4}$ inch in breadth. The cuttings were made sometimes from one side only, but generally they are scored all over the tree, about half of its bark being thus removed. The method of making the balls of rubber, which average two inches in diameter, is as follows :—A quantity of milk is dabbed upon the fore-arm, and being peeled off forms a nucleus. This is applied to one after another of the fresh cuts, and being turned with a rotary motion, the exuding milk is wound off like silk from a cocoon. The affinity of this liquid for the coagulated rubber is so great that not only is every particle cleanly removed from the cutting, but also a large quantity of semi-coagulated milk is drawn away from beneath the uncut bark, and during the process a break in the thread rarely occurs. By working hard, one person can collect five pounds of rubber per diem, though the average is only half the amount. I was assured, however, that in the interior where the trees run large it is no uncommon thing for one man to collect seven, or even nine, pounds in a day."

In the districts of Mungao and Kilwa alone the de-

velopment of India-rubber has created a new trade, which
gives a profitable employment to those classes whose means
of subsistence ceased with the suppression of the illegal
slave trade. The total export from the above-named
places exceeded, in 1880, 1,000 tons. The price also rose
rapidly at this period from £140 to £250 per ton. In a
report on the trade of Mozambique for the year 1880,
Consul O'Neill says : "It is curious to note the marvel-
lously rapid development of the India-rubber industry. In
1873 only £443 worth of India-rubber passed through the
Custom House of Mozambique. In 1876 it reached the
value of £22,198, and in 1879 it exceeded £50,000. It
would seem now almost to have reached its climax, while
the present rude method of collecting this produce prevails,
and until communications with the interior are properly
opened up, for the careless cutting of the tree by the un-
taught hands of the natives has resulted in the destruction
of enormous tracts of India-rubber forest near the coast."

L. Petersiana.—This is noted in the "Kew Report" for
1880 as a new species, furnishing a third variety or quality
of rubber on the East Coast. The plant was described as
growing near Tanga, on the coast of the mainland opposite
Pemba. The mode of preparing the rubber is said to be
quite different from that practised with the other kinds,
inasmuch as the juice is collected in a fluid state by tapping
and then coagulating it by heat. The quality of the rubber,
however, is said to be inferior to that of the other species.

To Sir John Kirk is due the credit of having introduced
to the Royal Gardens, Kew, living plants of four species of
Zanzibar rubber-yielding *Landolphias*, from which supplies
have been sent to the British colonies. From a return of
the imports of caoutchouc into London and Liverpool during
the year 1889 it seems that 5,919 tons were imported,
against 5,080 tons in 1888.

BORNEO RUBBER.—A good deal of uncertainty has always

been attached to the sources of rubbers or caoutchoucs from the East. From specimens brought to Kew, however, in 1880, it would seem that all the rubber from the Malay Peninsula is furnished by species of *Willughbeia* and *Leuconotis*, allied genera belonging to the natural order Apocynaceæ, to which the *Landolphias* before mentioned as the sources of African rubber belong. The following species of *Willughbeia* have been referred to in the "Kew Report" for 1880 as the probable source of the Malayan rubbers :—

1. *Willughbeia Burbidgei*, nov. sp., known as Manungan pulau.

2. *W. Treacheri*, nov. sp., known as Bertabu, and probably the source of the rubber of North-West Borneo.

3. *Leuconotis eugenifolius*, known as Manungan bujok, which is said to yield the best gutta (or rubber) of the Bornean woods.

4. *Chilocarpus viridis*, also a Bornean species of Apocynaceæ yielding caoutchouc.

5. *Chilocarpus flavescens*, nov. sp. Of this, excellent specimens were received at Kew in 1880 from the Singapore Botanical Gardens. The quality of the rubber from this plant was reported upon by Mr. Silver as "very fair" and "useful in our manufactures." It was valued at the time at 1s. 3d. a pound.

From the great attention that rubber-yielding plants have received at Kew during the past few years, it would seem to be proved that what is known in commerce as Borneo rubber is not yielded by *Urceola elastica*, as has been generally stated in all works hitherto dealing with the subject, but by one of the plants here mentioned, or at least by some allied species.

FIJI RUBBER (*Alstonia plumosa*).—A sample of this was first sent to Kew in 1877 by Sir Arthur Gordon, the then Governor. It was favourably reported upon as a "strong, elastic, pure rubber of the same character as the higher grades of African rubber." It proved to be the produce of *Alstonia plumosa*, another apocynaceous plant.

LIBERIAN RUBBER *(Ficus Vogelii)*.—This species was introduced in 1878 by Mr. Thomas Christy from Liberia. It was discovered by Vogel at Grand Bassa, and described in the Niger flora under the goneric name of *Urostigma*, which is now united with the genus *Ficus*. The rubber, which was formerly made into balls about the size of a large orange, has been valued in the London market at 1s. 6d. per pound, and would fetch a higher price if sent home in a cleaner state.

The trees yielding this rubber are known in West Africa by the name of "Abba," and in the *Kew Bulletin* for November, 1888, and May, 1890, the subject is discussed in detail. The rubber is described as containing a large amount of resin, which has prevented its general use in this country. Some improvements, however, have recently been made in collecting the fresh milk, so that a better quality of rubber has been obtained. In a note from the District Commissioner at Badagry to the Colonial Secretary at Lagos, published in the *Kew Bulletin* for May, 1890, referred to above, the coagulation of the milk is thus described :—

"When the milk is first brought to me in gin-bottles, I at once strain it into perfectly clean bottles through a piece of muslin fixed in a frame. The bottles are then allowed to stand for twenty-four hours for the milk to rise. It is then poured into a large tin, and put on the fire to boil. If much water is seen with the milk, none is added ; but if only a little, about a pint of water is added to every six bottles. As the water and milk begin to boil, lime juice is added in the quantity of one lime to each bottle. This assists the rubber to coagulate. When all the rubber in the water has formed into a large lump, it is taken out and forced into the moulds perforated and fixed in wood cases. Heavy weights are then laid on for twelve or twenty-four hours, and then the rubber is taken out, when it will be found ready for shipment. If one could only induce the natives to collect the milk, a large trade might be done ; but they are intolerably lazy, and do not care to attempt a new trade."

The report upon this rubber from the India-Rubber, Gutta-Percha, and Telegraph Works Company at Silvertown, dated March 20th, 1890, was fairly satisfactory. With a careful system of collecting and preparation from the plant, it is anticipated that the rubber may become a useful addition to our African supplies.

Regarding the growth of the plant, and the probability of its increased cultivation, it is stated in the *Kew Bulletin* that " the Abba trees of West Africa are widely distributed, and are generally used as shade-trees in market places, streets, and compounds. They can be propagated by ' the simple method of cutting off the branch and pushing it into the ground; and on account of the facility and rapidity with which it grows, the natives use it largely for fence posts.' " It is further stated : " From the trees already in full growth in the bush and towns, a considerable export trade could be rapidly established, and systematic planting would develop the trade to almost an unlimited extent."

MACWARRIEBALLI, OR DEMERARA RUBBER (*Forsteronia gracilis*).—This was first brought to notice in the *Kew Bulletin*, No. 15, for March, 1888, and is the produce of a large twining plant belonging to the natural order Apocynaceæ. From a report on this rubber obtained from the India-rubber, Gutta-percha, and Telegraph Works Company at Silvertown, by the Kew authorities, it would seem that if the substance can be obtained in any quantity it may eventually become a commercial article.

JAMAICA RUBBER (*Forsteronia floribunda*).—An allied plant to the former, and having a similar habit; it is confined to the Island of Jamaica, where it is known as Milk Wythe or Milk Vine. The rubber was first brought to notice in

the Report of the Director of the Botanical Department of
Jamaica in 1883, and again in 1884, and the whole subject
is fully treated of, together with reports on its probable
usefulness, in the *Kew Bulletin*, No. 24, for December, 1888;
from this report it seems that the rubber will probably
prove to be of considerable commercial value, and the notice
thus taken of it may perhaps lead to an extensive cultivation of the plant.

Other rubber-yielding plants of minor importance have
attracted the attention of the Kew authorities during the
past ten years, as *Urceola esculenta*, at one time known as
Chavannesia esculenta, first noticed as a rubber-yielding
plant by Mason, in Burma, in 1860; *Willughbeia edulis* or
W. martabanica, a Malayan plant; and *Chonemorpha macrophylla*, a large scandent evergreen shrub of the Andamans,
which is stated to yield a considerable quantity of caoutchouc.

We have gone thus fully into the sources of caoutchouc
in consequence of its very great and increasing commercial
importance. As a proof of this we may quote from the
annual report for 1886 of one of the principal brokers in
the trade, that "The world's consumption of all kinds is
steadily increasing," and we may further quote the following
statistics of the imports and value of raw and manufactured
rubber for the past seven and five years respectively :—

RAW CAOUTCHOUC.

1883	-	-	- 229,101 cwt.	value	£3,652,817
1884	-	-	- 198,844 ,,	,,	2,272,499
1885	-	-	- 180,141 ,,	,,	1,981,735
1886	-	-	- 192,518 ,,	,,	2,202,746
1887	-	-	- 235,539 ,,	,,	2,682,545
1888	-	-	- 218,171 ,,	,,	2,529,436
1889	-	-	- 236,274 ,,	,,	2,612,704

MANUFACTURED CAOUTCHOUC.

1882	-	-	- 1,447,739 lb.	value	£154,924
1883	-	-	- 2,073,374 ,,	,,	211,408
1884	-	-	- 2,612,740 ,,	,,	262,336
1885	-	-	- 3,139,632 ,,	,,	397,730
1886	-	-	- 2,681,210 ,,	,,	353,729

CHAPTER II.

GUTTA-PERCHA.

THIS product ranks next in importance to India-rubber, and although somewhat similar in substance and chemical properties, is tough and inelastic, which causes it to be inapplicable for many purposes to which rubber is put. Gutta-percha when exposed to the air and light absorbs oxygen and changes into a brittle resinoid substance, which unfits it for exposed uses. Under water, however, or in the dark it does not readily change, and lasts for a considerable period.

The history of the discovery and development of gutta-percha is one of great interest. It was first brought to notice in 1842, at Singapore, when it attracted much atten tion, and soon found its way to Europe; one of the first uses to which it was put being for soling boots, in consequence of its imperviousness to water, and its supposed greater durability than leather. Being easily moulded by heat, it was soon applied to the manufacture of pails, buckets, basins, water-pipes, door-handles, knobs for drawers, and a host of similar purposes. In consequence of its being a non-conductor of electricity, coupled with its durability under water, it soon became used for coating the wires of deep-sea telegraphs; for this purpose, however, India-rubber is now much more extensively used. The plant was originally described by Sir W. J. Hooker, in 1847, in the *Journal of Botany*, under the name of *Isonandra Gutta*, which has since been sunk under that of *Dichopsis Gutta*, Benth., by which name the true gutta-percha-yielding tree is now known. It is a large tree, sixty to seventy feet high, with a trunk two to three feet in diameter. At the time of its discovery it was abundant at Singapore, but during the

next five or six years it was so persistently destroyed for
the extraction of the milky juice that the tree was almost
exterminated, and at the present time only a few trees, that
are carefully preserved as curiosities, exist at Singapore.
In 1847 it was plentiful at Penang, but a similar fate has
overtaken it there.

To collect the milk the trees are cut down, and the bark

GUTTA-PERCHA (*Dichopsis gutta*).

stripped off, when it flows readily, and is collected in a
cocoa-nut shell, the spathe of a palm, or some similarly im-
provised vessel, and formed into blocks or lumps of various
sizes and shapes, the fluid quickly coagulating on exposure
to the air. The average quantity obtained from one tree is
about twenty pounds, and as the imports into this country
amount to between 40,000 and 60,000 cwt. annually, an
enormous number of trees have to be sacrificed to supply
the demand. In consequence of this destruction, and fear
of the entire loss of the article to commerce, attention was

drawn to it in the Kew Reports for 1876 and 1877, which resulted in the Government of the Straits Settlements taking up the question, so that in 1878 Dr. Dennys, Assistant Curator of the Raffles Museum, Singapore, drew up an important report to that Government, some extracts from which it may be both useful and interesting to quote. The true gutta-percha from *Dichopsis Gutta* is known in the Straits Settlements as gutta-taban. " It does not appear that the juice is collected at any special period. Mr. Lowe states, however, that there is a very marked difference in the yield of the wet and dry seasons; at the former period an average tree will yield some five catties (a catty $= 1\frac{1}{3}$ lb.), while in the dry season it will only yield one. Considerable difficulty, by the way, appears to exist in ascertaining the actual yield per tree ; and the difficulty will, owing to native habits of exaggeration, continue until some trustworthy European himself watches the operation. Mr. Murton states that a native gutta-percha merchant mentioned forty catties as the yield of a single tree ; while he himself, from other information, puts down the yield at from five to fifteen catties per tree, and never exceeding twenty. . . . . In view of the enormous number of trees which must have been destroyed if even ten catties be taken as an average, I should be inclined to accept the higher estimate.

"The destruction of trees involved in the process of collection is so enormous that it seems impossible for the supply to long continue. It is computed that over 7,000 trees were cut down during 1877 in the neighbourhood of Klang, while 4,000 must have perished near Selangor in a single month to furnish the 270 piculs (a picul $= 138\frac{1}{3}$lb.) returned as exported. The estimated annual export from the Straits Settlement and the Peninsula was given as ten millions of pounds in 1875, which, at the high average of fifteen pounds to a single tree, would give 600,000 trees. The demand seems always to exceed the supply.

"The principal adulterant made use of seems to be gutta-jelutong.

"Singapore and Penang are the chief collecting depôts for gutta-percha, and a failure in the supply might seriously injure the trade of either port."

The gutta-jelutong referred to here is the milky juice of an apocynaceous plant described by Sir Joseph Hooker in the *Journal of the Linnean Society*, vol. xix., p. 291, and also in the *Flora of British India*, vol. iii., p. 664, under the new generic name of *Dyera* (in honour of Mr. W. T. Thiselton Dyer, Director of the Royal Gardens, Kew), and the specific name of *costulata*.

Next in value to true gutta-percha is gutta-sundek, which is obtained in large quantities from Perak. Of this tree Dr. Trimen, of the Botanic Gardens, Peradeniya, Ceylon, reports in 1880 :—"I have during the year, through the kind exertions of Mr. Low, our Resident at Perak, received a consignment of germinating seeds of the second best variety of that country. This is called gatah-sundek, and Mr. Low informs me that it forms a very large tree, 120 feet high, but quick-growing. From specimens of the foliage and fruit sent with the seeds, it would appear (so far as it can be identified without flowers) to be a species of *Payena*. This is a valuable gift, as the gatah trees in Perak sufficiently large to produce the gum are now very rare, and very great difficulty arises in procuring seeds or specimens." Dr. Trimen further reports that the young plants were growing vigorously in Peradeniya and Heneratgoda.

Referring to the aid that the Colonial Government might give in preserving these valuable trees or extending their growth, Dr. Dennys says :—"It may be difficult for the Colonial Government to exercise a direct influence in favour of care and prudence on the part of the native administrations, but much might be done to encourage enterprise in the formation of new gutta plantations. It

may also be worth while to ascertain whether the appoint-
ment of European conservators, under the control of the
residents, would not achieve the end of preserving a most
valuable monopoly to the different governments; as it may
be assumed that the expenses thus incurred would be amply
justified by the commercial results, both to Singapore and
Penang as depôts, as well as to the original collectors and
vendors of such important articles of trade.

"It is not impossible, also, that fresh discoveries might
be made, if not of new trees yielding similar products, of
sub-varieties which might furnish a commercially valuable
substitute; while it is more than probable that vast areas
of virgin growth might be discovered in the interior portions
of the Peninsula by an explorer under Government auspices.

"The principal obstacles in the way of individual enterprise lie in the time necessary to mature the tree, said to
be about fifteen, or perhaps twenty years at least, and the
difficulty of obtaining seeds, saplings, or cuttings wherewith
to commence plantations. These can only be met by the
cordial co-operation of the residents and native authorities,
the latter especially needing to be convinced that by aiding
the movement they will not be depriving themselves of a
valuable monopoly. As regards the former, it is probable
that but very few Europeans would embark capital which
would not yield an outturn for fifteen or twenty years,
which I am informed, on botanical authority, is the average
time required before a tree is ready for tapping; many
trees, indeed, are reputed to be thirty years old when
tapped, and it would therefore seem that the Government
alone can afford to undertake the establishment of plantations. At present we are without data as to probable
expense, but as the trees are essentially jungle trees, and
require no care when once fairly started, this may be taken
as very low. Assuming each picul of 133⅓ lb. of the best
qualities to represent the yield of ten trees, and to be

worth $45, 10,000 trees would give a gross return of
$45,000. The available Crown lands in Singapore could
probably grow 100,000 trees, at the lowest estimate, giving
$450,000 in the gross outturn, though this estimate must be
mere guesswork until a proper survey be made. But
assuming that the annual income of the colony could be
increased by $200,000, or less than half the sum named,
the matter seems worth attention; while there is reason to
believe that even if the yield from the native states continued at its present figure, the additional supply would
soon find a market without materially lowering the price."

Though we have referred here to only two sources of
gutta—namely, gutta-percha or gutta-taban from *Dichopsis
Gutta*, and gutta-sundek from *Payena (Ceratophorus) Leerii*
—it is very clear that the substance is yielded in the east
by other allied trees belonging to the Sapotaceæ; and
though but little or nothing is known of them scientifically,
it is very important, both for the sake of science and commerce, that some steps should be taken to obtain materials for
a complete investigation of this interesting question. Though
the complete history of the substance is encompassed by a
period of forty years, its importance to the commerce both of
this country and our eastern possessions is immense; and
there is no reason why, with the discovery of new sources of
produce, or the proper care and development of the old
sources, the trade should not still be considerably extended.

The imports and value of gutta-percha into this country
have been declining for the past few years until last year,
when they suddenly rose, as will be seen from the following
returns since 1884 :—

1884	- - - 62,713 cwt.	- - -	value	£462,745
1885	- - - 53,894 ,,	- - -	,,	348,104
1886	- - - 40,697 ,,	- - -	,,	269,808
1887	- - - 24,145 ,,	- - -	,,	156,563
1888	- - - 22,483 ,,	- - -	,,	181,660
1889	- - - 48,042 ,,	- - -	,,	576,896

BALATA (*Mimusops globosa*)

A substance known as Balata, very similar to gutta-percha, but with the recommendation of its being more ductile, and consequently more durable, is obtained from the trunk of a large forest-tree sixty to seventy or even one hundred feet high, said to range from Jamaica and Trinidad to Venezuela and French Guiana. It belongs to the natural order Sapotaceæ, to which the other gutta-percha-yielding trees belong, and is known to botanists as *Mimusops globosa*. Balata was first brought to notice in this country in 1859, at the instigation of Messrs. Silver and Co., of London, Dr. Van Holt, of Berbice, having previously noticed the presence of the substance in the bark.

The sample sent to Messrs. Silver by Mr. David Melville did not prove, upon examination, so satisfactory as was anticipated, and nothing more was heard of the substance till the International Exhibition in London in 1862, where specimens were exhibited in the British Guiana court, and Sir William Holmes brought it into prominent notice. "The result of his zealous efforts was that Messrs. Silver and Co. applied for a further sample, and some of the dried material was sent them. A better opinion of the merits of balata seems to have been derived from this experiment. Some appears, also, to have been submitted about the same time to the Gutta-percha Company, in London, and a demand was created for it. In 1865, three years later, the quantity exported was 20,000 lb. Then the trade commenced to fall off, and continued to decrease till 1874, when the amount sold only realised £111. In 1877 the demand revived, but fell again the following year; reviving again the next year, and increasing up to 1884, when there was a falling off."

In the ten years between 1875 and 1885 the India-rubber, Gutta-percha, and Telegraph Works Company, at Silvertown, used balata to a great extent, but since the

D

latter date it has been given up by them, in consequence of its having been found not durable when exposed to the air. Articles manufactured from it are said to have cracked on the surface, and the inner portion to have lost its tenacity. Balata seems to be unsuited for mixing with gutta-percha, and has, therefore, been generally used alone. At the present time it is estimated that not more than fifty tons are imported into this country annually, and that probably another fifty tons go to France and the United States.

As imported, balata, as a rule, is very uniform in quality, fetching about 2s. per pound, while gutta-percha ranges between 1s. and 3s. 6d. per pound, according to quality.

GUTTA-SHEA (*Butyrospermum Parkii*).

Under this name a substance has been brought to notice during the past few years which resembles gutta-percha in many respects, but is more brittle. It is obtained from shea butter, the solid oil or fat expressed from the seeds of *Butyrospermum Parkii*. The tree is a native of Western Africa; and the fat is not only used by the natives as butter, but it has also been exported in considerable quantities to this country, since 1851, for soap-making. It is estimated that the quantity annually imported from Sierra Leone amounts to about 500 tons. It is used in this country only for hard soaps; and in the course of manufacture gutta-shea is found to be present to the extent of from ·5 to 75 per cent. On the Continent it is largely used for candle-making. In noticing this product, in the Kew Report for 1878, Sir Joseph·Hooker says, " It is insoluble in alcohol, a mixture of alcohol and ether, acids, and alkalies. It is slightly soluble in pure ether, and ordinary animal and vegetable oils and fat. From the extremely small proportion in which it is present in shea butter, its extraction would not be profitable; and regarded as a bye-product it does not appear to be suitable for any purpose to which gutta-percha itself is applied." Gutta-shea is, therefore, a substance which, though an introduction of recent years, does not, according to present lights, appear to have much of a future before it. Nevertheless, some new application may before long be found for it, and it may yet become an important trade product.

MASSERANDUBA or COW-TREE OF PARA.

(*Mimusops elata.*)

A tree one hundred feet high, belonging to the same natural order (Sapotaceæ) as the gutta-percha, balata, and gutta shea trees. Though attributed, as above stated, to *Mimusops elata*, its specific name has not been determined with certainty, as the flowers have not been examined. The milky juice flows freely from wounds made in the bark : it is of a cream-like appearance and substance, thickening and becoming like gutta-percha on exposure to the air. This substance was first brought to notice by the South American traveller, Richard Spruce, who sent a sample of the wood and milk, collected by himself, to the Kew Museum in 1849, where they are still preserved.

Up to the present time the milk of the Masseranduba has not been turned to any practical account.

GUM EUPHORBIUM.

Under this name, or more frequently under the contraction of "G. E.," by which it is known in commerce, a substance has quite recently attracted much interest, in consequence of its reputed adaptability for mixing with gutta-percha and india-rubber to the extent of fifty per cent. It was stated in September, 1887, that a piece of vulcanised rubber containing fifty per cent. of euphorbium gum was tested for some time in an exposed position on a roof, and that it kept in better condition than a similarly exposed

D 2

piece of ordinary pure vulcanised rubber. In combination
with gutta-percha it is said to reduce the tendency to become
brittle. Washers made with thirty per cent. of this gum
and vulcanised stand well, and retain their elasticity. The
gum comes to this country in small balls or ovoid masses
from one to four ounces in weight each ; in fracture it has
the appearance of broken balls of putty, friable when cold,
but easily kneaded or pulled out in hot water. It is very
clean, and has scarcely any admixture of bark or other
impurities. The bulk of the gum is said to come from the
Cameroons district. Nothing definitely is known of its
botanical origin, except that it has been referred by some to
a species of *Euphorbia*, and by others to a species of
Tragia.

A concrete milky juice, identical with this gum, was imported into Liverpool in 1874 from St. Paul de Loanda, and
was then referred to the genus *Euphorbia*. Another sample
of a similar character was received at the Kew Museum in
1883 from Mossamedes, and upon being tried by a well-
known firm of linoleum manufacturers, was reported to be
likely to prove of some value in their trade. The development of this gum, as well as its botanical origin, is still in
the future.

CHAPTER III.

FOOD PRODUCTS.

The Englishman is so conservative in the matter of food that it takes a considerable time even to induce him to give a trial to any article with which he is unacquainted, and a much longer period before he will accept it as a regular, or even an occasional, contribution to his diet. What he eats when travelling on the Continent, or in foreign lands, his soul abhors when he returns to his native shores ; therefore new food products are not by any means abundant, and if, perchance, they do occasionally appear in our markets, they are looked upon only as a novelty to be recorded in some journal, and again lost sight of, at least for a time.

The greatest strides made in the commerce of food products during the present century have been in the development of already existing sources. It is not within our scope to trace the progress of these articles, but we may, perhaps, legitimately, but briefly, point out to what extent steam has assisted in placing within our reach foreign and colonial produce of vegetable origin. Take, for example, the increased and ever-increasing quantities of fresh and preserved fruits, the latter hermetically sealed in tins, that are now poured into our markets at all seasons of the year, by which means our tables are furnished with fruits all the year round that but a very few years since we could enjoy only at certain seasons. The pine-apple is one illustration of this; fresh fruits being brought from the West Indies by fast-going vessels, and the preserved whole fruits in tins from Singapore, Bahamas, and Natal, the flavour of which is almost, if not quite, equal to that of the fresh fruits. But a very short time since, tinned pine-apples were not known to appear on the tables of even the well-to-do middle class, but the excellence of their quality has won for them a place at the present time amongst the more costly fruits on the tables of the upper classes.

For some years past we have received consignments of tinned fruits, such as apples, pears, peaches, &c., from America and Australia, but it is not till within the last five or six years that anything like a real and earnest attention has been given to the subject of the general exportation from our colonies into the mother country of preserved and fresh fruits. A great impetus was given in this direction by the Colonial and Indian Exhibition of 1886, not only by the extensive and various exhibits of colonial fruits, but by the establishing of a market for the sale of the fresh produce brought from America, the West Indies, and even Australia, by fast-going steamers; by this means, though the supply was intermittent, some of the fruits both of tropical and sub-tropical countries were brought under the notice, and within the reach of many who had never probably even seen them before. As this question of the development of the trade in colonial fruits formed the subject of a special report in connection with the Exhibition, by Mr. D. Morris,* it will suffice to give a few notes from that report, and to point out that the entire subject therein dealt with is one that has originated and grown to its present proportions within the last twenty years. Regarding Canadian fruits, Mr. Morris draws attention to the enormous proportion which the apple trade has assumed; and states that the province of Ontario alone exported half a million barrels, all of which, however, did not come to England, a considerable quantity going direct to Norway and Denmark. In the matter of packing, always a difficulty with fresh fruits, it is stated that " nothing was, on the whole, so satisfactory as a

* *Fruits*, by D. Morris, M.A., F.L.S., Colonial and Indian Exhibition, 1886. *Reports on the Colonial Sections of the Exhibition.* Edited by H. Trueman Wood, M.A. London : Clowes and Son.

careful wrapping of fruit in tissue paper to prevent bruising, and giving them as much air as possible in a cool storage."

Apples, pears, oranges, lemons, and a great variety of well-known fruits came from the Australian colonies and New Zealand ; while from the Cape and Natal came similar fruits, as well as many others of indigenous growth, such as the AMATUNGULU or NATAL PLUM (*Carissa grandiflora*), the KEI APPLE (*Aberia Caffra*), and others. Though both of these fruits have a pleasant acid and refreshing flavour, and have been introduced to notice before, so long back, indeed, as the International Exhibition of 1862, they are not yet known in this country, except as curiosities ; the same may be said of many of the Eastern fruits, such as the RAMBUTAN (*Nephelium lappaceum*), LOVI LOVI (*Flacourtia inermis*), the CARAMBOLA (*Averrhoa Carambola*), the MANGO (*Mangifera indica*), and a large number from the West Indies, including the CHERIMOYER (*Anona cherimolia*), the SWEET SOP (*Anona squamosa*), the GINEP (*Melicocca bijuga*), the SWEET CUP (*Passiflora edulis*), the NASEBERRY (*Achras Sapota*). These and many others have been introduced to notice during the last fifty years, but have never found their way into English commerce as regular articles of trade, either in a fresh or preserved state. We may, however, hope, and are led to expect, that with the interest awakened in the subject during the past year or two, a new and profitable branch of commerce in this direction may be established ere long between the British colonies and the mother country ; and as an example of the benefit to be thus derived. we may refer to the case of Fiji, one of the newest of our colonies, where the value of the exports of fruit to Australia has increased from £507 in 1877 to £23,994 in 1884. We have thus far referred to those fruits that were only occasionally brought to us from distant parts of Greater Britain. As an illustration of the progress of the regular import trade in certain fruits, we will quote the following table, as given by

Mr. Morris, which shows the increased values during the forty years ending in 1885 :—

	1845	1865	1885
Apples, Oranges, Lemons, &c. -	£158,098	£1,131,183	£3,619,788
Nuts, Almonds, &c. - - - - -	80,682	424,866	701,910
Currants, Raisins, Figs, &c. - -	648,108	1,629,935	3,265,825
	£886,888	£3,185,984	£7,587,523

Amongst home-cultivated fruits none have increased of

TOMATO (*Lycopersicum esculentum*).

late years more rapidly, both in favour and in extended culture in this country, than the TOMATO (*Lycopersicum esculentum*). Originally a native of Mexico or South America, the plant has spread by cultivation into tropical and temperate climates, and in England its cultivation has been gradually extending during the past thirty years, until the consumption has increased to such great proportions that it has recently been estimated at sixty pounds per head of the whole population. So great is the demand, indeed, that to meet it enormous quantities are brought here from Jersey, France, and other parts of the Continent; and when the season for fresh tomatoes has passed, the tinned fruits from America are ready to take their place.

Tomatoes are no longer used only for making sauce or for pickling, the fresh fruits boiled down with sugar make an excellent preserve, scarcely to be distinguished from apricot. Simply boiled in water they are a palatable vegetable, besides which they may be cooked in a variety of other ways, and they make excellent salads, while the tinned

CUCUMBER (*Cucumis sativus*).

fruits likewise lend themselves to many forms of culinary treatment.

The CUCUMBER (*Cucumis sativus*) is another fruit the cultivation and consumption of which has extended very much during the past few years. Though it has been grown in England from a very early period, it was not generally cultivated till the middle of the fifteenth century, and it is within comparatively recent years that its growth has assumed the proportions we now see.

Besides frame cucumbers of many varieties, grown especially for table use, the plant is cultivated as a field crop, covering extensive areas, particularly in the home counties,

whence the fruits reach the London market, and are largely
used for pickling.   It has been stated that " in the market-
garden district of Sandy, in Bedfordshire, ten thousand
bushels of pickling cucumbers have been sent away in one
week."   Cucumber cultivation, therefore, represents a large
sum of money in the aggregate.   In the early season they
realise high prices, but in August and September they are
not unfrequently sold at the rate of a penny a dozen.

     This account of cultivated plants would scarcely be com-
plete without some notice of the extension of the POTATO
(*Solanum tuberosum*) cultivation in England, and though much
of this was effected before the period of which we have to treat,
nevertheless a very large number of the best recognised varieties of the potato are of recent origin.   When we consider to
what perfection cultivation has brought this vegetable, which
in its native Chilian form is a globular waxy tuber not larger
than a walnut, and of which a well-known writer prophesied
in 1708 " that it might prove good for swine," we may
perhaps be surprised that even more has not been done
during the last fifty years in introducing and establishing
entirely new sources of food, such, for instance, as allied
species of *Solanum*, that might under cultivation produce
edible tubers.   Something in this direction has indeed been
done quite recently by Mr. J. G. Baker, F.R.S., who, in a
paper on " Tuber-bearing Species of Solanum," read before
the Linnean Society in 1884, and published in the Journal
of that Society, vol. xx., p. 489, pointed out that *Solanum
Maglia*—a close ally to *S. tuberosum*, and like it a native of
Chili—produced similar tubers.   The plant, indeed, seems
to have been introduced to the Royal Horticultural Society's
garden at Chiswick in 1822 and to have been mistaken for
the true potato.   *S. Commersoni*, a native of Uruguay,
Buenos Ayres, and the Argentine Republic, also produces
tubers, as well as *S. Jamesii* from the mountains of the
South-west United States and Mexico.

Mr. Baker suggested " that these should be brought into the economic arena and thoroughly tested as regards their economic value, both as distinct types and when hybridised with the innumerable *tuberosum* forms." Trials at hybridisation were subsequently made at Messrs. Sutton's grounds at Reading, under the superintendence of Earl Cathcart, Mr. Baker, and Mr. A. Sutton, with the result that when

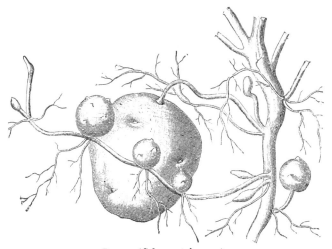

POTATO (*Solanum tuberosum*).

the tubers were dug up those of *S. Maglia* were found to be as large as ordinary potatoes and of fairly good quality when cooked, while the other species were not so satisfactory, so that in *S. Maglia* we have probably a new addition to our future supplies of edible tubers.

Amongst tuberous roots that have been recommended at different times during the present century, and even tried as a substitute for the potato, may be mentioned the ARRA-CACHA (*Arracacia esculenta*) an umbelliferous plant, the

native country of which is not accurately known, though it
is now generally cultivated as a food-plant in Venezuela,
New Granada, and Ecuador. The plant was first introduced
to Kew between fifty and sixty years since, but it was found
to be unsuited for outdoor cultivation ; since then the roots
have been frequently re-introduced, but further trials have
proved that the English summer is not sufficiently long to
mature them. They have, however, been recently introduced
both into the West and East Indies, and promise to become
established food-plants in both countries. Another plant
that attracted some attention during the early years of the
potato murrain as a suggested substitute was the *Ullucus
tuberosus*, an herbaceous plant belonging to the natural order
Chenopodiaceæ, a native of Peru and Bolivia, where it is
extensively cultivated for the sake of its tubers, which are
largely eaten as food. In appearance they are very like
potatoes, in their wild state about the size of hazel nuts, but
as cultivated at Kew in 1851 they grew to about two inches
long and one inch thick. The substance of the tubers is
close and waxy, and this was not improved by cultivation,
consequently they did not meet with approval.

Under the name of CHOCO or CHAYOTE the fruits of
Sechium edule, a climbing plant belonging to the Cucurbitaceæ, and native of Tropical America, whence it has been
introduced to Madeira and the Atlantic Islands, is occasionally seen in the London market. They are eaten simply
boiled, and are said to somewhat resemble a vegetable
marrow. They were first introduced thirty or forty years
since, but have never found much favour here. Within the
last two years the plants have been introduced into Ceylon,
India, and Singapore, where they seem to thrive and produce
a plentiful supply of fruits.

Two of the most recent introductions to cultivated vegetables are the SHANTUNG CABBAGE (*Brassica chinensis*) and
the CROSNES (*Stachys Sieboldi*, better known as *S. tuberifera*).

The first of these is fully described in the *Kew Bulletin*, No. 17, for May, 1888, from which we gather the following notes : Seeds of this cabbage were offered to the Royal Gardens, Kew, in April, 1887, by a correspondent who had occupied an important position at Chefoo, China. He says, " I have just received from Chefoo, North China, a small packet of Shantung cabbage seed, and I should like, if possible, to introduce this delicious cabbage into England. It grows in the north of China, is lettuce-shaped, and weighs from five to eight pounds. When boiled it is nearly as good, if not quite, as sea-kale ; eaten raw in a salad it is of so delicate a flavour that I know of no vegetable in England to approach it. It is an autumn cabbage, should be planted about eighteen inches apart, thrives best with moisture, and in Shantung is well watered every day ; there the seed is sown in June. When nearly full-grown it should be tied round so as to give it a good white heart. If it can be acclimatised in this country it will be a great addition to our vegetables." The seeds received at Kew " were sown in a heated pit on the 3rd of May, and in about a fortnight all had germinated ; they were pricked off into boxes, and when large enough transferred to pots. They were kept in a cold frame until the beginning of June, when they were planted out in beds of rich soil, about eighteen inches apart in the rows and the same distance from row to row. About the middle of July the plants were tied up in the same way as cos lettuce, and when well filled and blanched were cut for use. They were pronounced to be excellent. It is possible that this Chinese cabbage may prove a useful addition to English gardens."

The CROSNES (*Stachys Sieboldi*). The first notice of this new vegetable appeared in the *Revue Horticole*, 1885, p. 236, and in 1887 was introduced for use in France, having been cultivated by M. Paillieux, who now grows it on a large scale for the Paris market, where it is known under the above name of " Crosnes," from the locality where its culture is carried on. It was first referred to *Stachys affinis*, then to *S. tuberifera*, and now to *S. Sieboldi*. In December, 1887, tubers from plants grown at West Moulsey were ex-hibited at the Royal Horticultural Society of London, since which time it has been grown in many English gardens, and the tubers cooked and eaten, and from most of the growers they have been very favourably reported upon as an esculent, as well as for pickling in vinegar. We quote the following from an article on this new vegetable in the *Gardeners' Chronicle* for Jan. 7, 1888, p. 16 :—" The plant is alleged to be perfectly hardy, and of the easiest possible culture. It will grow anywhere, on any ordinary soil, but like other plants it will no doubt repay a little attention in the way of trenching and manuring. Its defects at present are its small size and the fact that its tubers do not keep well when lifted, but both these defects can be overcome or evaded. In point of flavour we have heard it compared to salsify, Jerusalem artichokes, and to boiled chestnuts ; our own taste would lead us to consider it as most nearly allied in point of flavour to the latter.

" M. Carriere, while admitting that the difficulty of pre-serving the tubers militates against them as a market-garden crop, points out that it is well suited for the domestic kitchen-garden, where the tubers can be lifted in late autumn or winter and eaten in a fresh state either boiled, fried like salsify, made into sauce, or cooked in a dozen other ways that any cook — especially a French one — will devise."

The plant belongs to the natural order Labiatæ, is a native of Japan, and produces its tubers abundantly at the ends of the underground branches, exactly as in the potato. It has been advertised by the seedsmen as " Chinese Arti-choke," but the name " Stachys " has been suggested for it as being more correct than either Chinese Artichoke or Crosnes.

In the matter of farinaceous foods, though no absolutely new source has to be recorded as the introduction of the nineteenth century, the extension of existing sources, both in supply and demand, has been remarkable. The most striking development, perhaps, is in CORNFLOUR, MAIZENA, and similar preparations of maize or Indian corn (*Zea Mays*). It is now nearly sixty years since William Cobbett so strongly advocated the growth of maize in this country in his treatise on "Cobbett's Corn," and even grew it extensively on his own farm at Nine Elms. It has since been proved, however, that though the plant will grow well with us, the average length of the English summer is not sufficient to thoroughly ripen the crops. Nevertheless, we import enormous quantities of the grain into this country from the United States ; the trade returns of which show that, while in 1856 444,453 cwt. of Indian corn or maize were imported into England, in 1889 the quantity had risen to 36,203,069 cwt., representing a value of £8,580,080 ; while of Indian corn meal, an article not known in 1856, there were in 1889 imported 24,066 cwt., valued at £19,365. Indian corn is not only an article of human diet, but it is also used for feeding horses, as well as in distilling.

RICE, the grain of *Oryza sativa*, is another article of a similar character which shows a large increase in imports and consumption of late years. Thus, from British India we received in 1889 4,632,097 cwt., and from other countries 1,950,652 cwt., making a total of 6,582,749 cwt., of the aggregate value of £2,689,363 ; while in 1856 the imports of rice from all countries were 3,692,001 cwt.

SAGO, TAPIOCA, and similar farinaceous food products, have also more or less increased.

Amongst pulses probably the most notable advance is in LENTILS, the seeds of *Lens esculenta*, a very ancient food-plant, cultivated by the Hebrews, and in Europe since the days of the Roman Empire, and at the present time throughout the East, as well as in many parts of Europe,

MAIZE (*Zea Mays*).

North Africa, West Asia, and North-West India.   Lentils
are used either whole as a vegetable, or ground into flour

for thickening soups, and are increasing in favour with us. Lentils, when deprived of their outer husk, are easily digested, and form an important part in the composition of the well-known foods so much advertised as " Ervalenta" and " Revalenta."

As an illustration of a neglected food product which might be made available, or at least largely developed, in this country, we may refer to a notice of chestnut flour in the " Kew Report " for 1879, where the fact is stated that the CHESTNUT (*Castanea sativa*) is an important article of subsistence in the Apennines, the chestnuts being

ORYZA SATIVA.

carefully dried and ground into flour and made into cakes. The flour having been analysed by Professor Church, he gave the following results :—

Moisture - - - - - -	14·0
Oil or fat - - - - - -	2·0
Proteids - - - - - -	8·5
Starch - - - - - -	29·2
Dextrin and soluble starch - - -	22·9
Sugar - - - - - -	17·5
Cellulose, etc. - - - - -	3·3
Ash - - - - - - -	2·6
	100·0

Professor Church further expressed an opinion " that chestnut flour ought to be of easy digestibility, and a

E

suitable children's food, considering that it contains over 40 per cent. of nutritious matters soluble in-pure water." Chestnut flour is now an article of commerce in this country.

We cannot dismiss the subject of food products without a brief reference to the continued increase in the imports and consumption of SUGAR. It is not within our province to follow the grievances of sugar planters, the question of bounties, and the fluctuations of the market; suffice it to say that when sugar can be bought retail at twopence per pound, it is truly within the reach of all. Though much of the sugar brought to this country is still furnished by the SUGAR CANE (*Saccharum officinarum*) (*see* Frontispiece), and is derived from the East and West Indies, Mauritius, and Brazil, the BEETROOT (*Beta vulgaris*) has become a formidable rival to the cane in consequence of the very great improvements that have been effected in the processes of clarifying and crystallisation, by which means the sugar is scarcely to be detected by an ordinary observer from cane-sugar. The cultivation of beet for sugar-making has been attempted several times in this country, but always on such a small scale that no practical result has ever come of it. When it was first introduced is not clear, but in 1837 a beetroot-sugar refinery was established at Chelsea, and many acres of land in and around Wandsworth and other parts of the suburbs of London were put under beet cultivation. It did not, however, succeed, and most of the land is now covered with houses. Beetroot culture has since been tried in Essex and other counties, as well as in Ireland, but without success. The supplies of beet for sugar-making purposes are chiefly obtained from Germany, France, Belgium, Holland, Austria, and Russia.

A very interesting subject in connection with the sugar-cane has been brought to light at Kew, and published in the *Kew Bulletin* for December, 1888, and October, 1889. The plant having been so long under cultivation, its native country, like that of many other cultivated plants, is entirely unknown. The sugar-cane being propagated by means of buds and suckers, the mature fruits have scarcely, if ever, been seen. It was reported, however, from Barbados, at the end of 1888, that seedling sugar-canes had been found, and were being raised at the Botanical Station in that island. It was further subsequently stated " that certain varieties of sugar-canes still retain the power of pro ducing mature seed." This is an interesting fact from a scientific point of view, but it is also of considerable interest to the sugar planters, who may be able to raise plants from seeds that may prove far superior to the old varieties both in quality of sugar and extent of yield.

Another kind of sugar which has attracted much attention is that obtained from the stems of the American Sorghum (*Sorghum saccharatum*), and hence called SORGHUM SUGAR. An extensive series of experiments in the cultivation of the several varieties of the plant, and the manufacture of sugar from the stalk, was conducted by the United States Department of Agriculture in 1879. At the same time, also, much attention was given to maize as a source of sugar. But these new fields of research have not at present affected our sugar supply, though it is not impossible they may ere long do so.

The total imports of sugar, refined and unrefined, in 1889 amounted to 26,567,505 cwt., against 7,761,240 cwt. in 1856.

E 2

CHAPTER IV.

BEVERAGES.

LIKE food products, the greatest progress made in the development of beverages of vegetable origin has been in the extension of applications, rather than the introduction of entirely new sources. The most marked extension of well-known beverages has been due to the introduction of the vine (*Vitis vinifera*) into Australia, the Cape, and Canada, all of which have become wine-producing countries, and the extension of the vine in each may be reckoned with the growth of the colony itself.

The first vineyard planted in South Australia dates back to the year 1840; in 1852, 282 acres of land were under vine cultivation, which in 1884 had extended to 4,590 acres, yielding 473,535 gallons of wine. Every year more land is being put under vine cultivation, so that the trade is still extending. In Victoria and New South Wales the vine is similarly spreading; a considerable amount of attention having been given to it, especially during the last twenty years. The ravages of the phylloxera in the European vineyards has indirectly helped to increase the demand for the colonial produce. The wine imports during the past thirty years show a considerable fluctuation: thus, in 1856 the returns were 9,481,880 gallons; in 1866, 15,321,028 gallons; in 1876, 19,979,838 gallons; in 1886, 14,561,913 gallons; and in 1889, 15,934,934 gallons. This falling-off since 1876 may, to some extent, be due to the spread of temperance principles, for we find that the imports of tea have increased from 86,200,414 pounds in 1856 to 221,602,660 pounds in 1889; while cocoa has also increased from 7,343,475 pounds in 1856 to 26,735,974 pounds in 1889. Coffee, on the other hand, shows a

diminution. One reason for the greatly increased consumption of tea must always be found in its very widely
extended cultivation, especially in British India, and still
more recently in Ceylon, from both of which countries we
now draw large supplies, indeed to the extent of nearly
one-half of the entire imports. These large and widened
resources have been the means of bringing tea within the

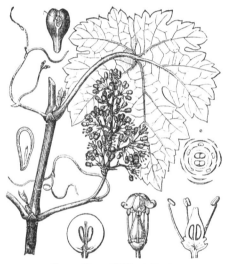

GRAPE VINE (*Vitis vinifera*).

reach of all classes of the community; and where good tea
often realised in the early years of the present century
eight or ten shillings a pound, it can now be had from two
shillings to two and sixpence.

Regarding cocoa, it is a comparatively recent development, and its use has made very rapid progress amongst us,
owing chiefly, no doubt, to the improvements that have
been effected in its preparation, so that instead of having to

obtain the broken seeds or "nibs," and boil them for half
an hour to obtain a cup of cocoa, the cocoa is now prepared
in a powdered and portable form, and the beverage can be
prepared with a minimum amount of trouble. These modes
of preparation, however, lend a ready means of adulteration
by the admixture of flour and similar substances, so that
the cocoa of the present day is not generally so pure a
beverage as when cocoa-nibs only were used. In view of
the increased demand for cocoa, steps were taken in 1880
by the Kew authorities to transmit from Trinidad some of
the best known varieties to Ceylon, Singapore, and Fiji.
Those sent to Ceylon and Singapore arrived in good order,
but those despatched to Fiji perished; a further consignment,
however, arrived in a healthy condition, and from all three
places favourable reports were received, so that cocoa has
now become established in countries far from its original
home.

On the subject of coffee little need be said, except to
recall the facts, so generally discussed at the time, of the
almost total destruction of the plants in Ceylon by the
Hemileia vastatrix. It will be fresh in the memory of all
how that in May, 1869, a few coffee plants were observed
to be attacked by a small fungus, which two months later
had spread over two or three acres, and after this worked
its devastation with extraordinary rapidity, till in 1872
there was not an estate in the island free from it, and it
also spread into Southern India ; in 1878 it appeared among
the coffee plantations of Java, in 1880 in Fiji, and in 1881
in Mauritius. Its ravages, continuing for several years in
Ceylon, had a most serious effect upon the plantations, and
many suggestions were made to rid the island of the pest ;
amongst others that of introducing an entirely new species
of coffee—namely, the Liberian sort (*Coffea liberica*)—
which it was thought would be disease-proof. In 1874–75
large quantities of seeds and plants of this species of coffee

were distributed from Kew to most of the coffee-growing countries, including the following :—Jamaica, Bahamas, Barbados, Bermuda, Dominica, Montserrat, New Grenada, Rio de Janeiro, Mauritius, Natal, Ceylon, Bangalore, Calcutta, Madras, and Java, from all of which subsequent reports were obtained of the entire suitability of the plant for cultivation, and recording its rapid growth and establishment. Liberian coffee, however, has not by any means displaced the older *Coffea arabica.*

Under the name of FEDEGOSA, NEGRO COFFEE, or CAFÉ DU SOUDAN, the seeds of *Cassia occidentalis* have been brought to notice as a substitute for coffee. The plant, which belongs to the Leguminosæ, yields its small seeds in great abundance, and these, when roasted, are used in the West Indies and Central America, as well as in Tropical Africa, where the plant has become naturalised, in the preparation of a beverage very similar to coffee. It is said, indeed, that after being roasted and ground they are often mixed with coffee as an adulterant. The seeds are further said to have tonic and febrifugal properties. They were introduced in 1876, but are now only occasionally seen in the European market.

PARAGUAY TEA or YERBA DE MATÉ.—Under these names the leaves of *Ilex paraguariensis* are extensively used in Paraguay and Brazil in the preparation of a beverage. Upwards of five millions of pounds are said to be exported annually from Paraguay alone. To prepare the tea for market, the leaves and twigs are simply gathered and dried in the sun or open air, after which they are broken or coarsely ground, from which an infusion is made. The tea contains caffein, and has been recommended at different times for use in this country. About four or five years since it was much advertised as a nervine tea, but it has since dropped out.

A very important non-intoxicating beverage, but of a

different class from those we have just been considering, is
LIME JUICE, obtained from the fruit of the lime (*Citrus
medica*, var. *acida*), a native of India, but now cultivated
on a very large scale at Montserrat.   The first lime-tree
orchards were planted in that island in 1852, and the
plantations of the Montserrat Company now cover an area

COFFEE (*Coffea arabica*).

of more than 600 acres, and contain 120,000 trees.   The
trees are systematically cultivated, and the juice prepared
especially for the English market for use as a beverage.   It
is brought to this country in large casks, and is here clarified
and bottled for the retail trade.   Mixed with water, it forms
not only an agreeable and refreshing beverage, but it is also

very useful in gout and rheumatism, as well as in fevers, indigestion, dyspepsia, etc. It is, however, best known as an antiscorbutic, and a considerable amount of attention has been directed to it since the return of the Arctic Expedition, as the outbreak of scurvy in the sledging expeditions was a serious drawback to their success. An inquiry into the matter resulted in the passing of an Act on August 20th, 1867, which rendered it compulsory, under certain penalties, for every ship to take on board lime or lemon juice in sufficient quantity to serve out so much per day to every member of the crew during the voyage. In consequence of this enactment, scurvy is now very little known in the English navy. The lime is cultivated on a small scale also in Dominica and Jamaica.

The utilisation of the kola or cola-nut in the preparation of cocoa is described under DRUGS.

CHAPTER V.

DRUGS.

THOUGH drugs or medicinal agents in use by foreign nations have always had a considerable amount of attraction for travellers, the consequence of which has been the frequent introduction from time to time of new remedies, it is during the last thirty years that the greatest achievements have been attained, not only in the discovery and introduction into this country of new medicinal agents of vegetable origin, but also for the introduction into our own colonies of others whose established reputation has caused them to become absolute necessities.

With the advance of chemical and medical knowledge, the herbs of our own gardens which were so highly prized for their very many supposed virtues by our grandfathers and grandmothers have rapidly fallen into disuse, and at the present time but very few British plants are included amongst those furnishing useful drugs. No better illustration can be had of the changes effected by the agency of rapid and cheap communication with the various parts of the world than a reference to the pages of the *Pharmaceutical Journal* for the last twenty years, or to the several numbers of Christy's *New Commercial Plants and Drugs.*

In 1837 there appeared two volumes, consisting of over 900 pages, treating of the medicinal plants of Great Britain, under the title of "The British Flora Medica." A new edition of this work was issued in 1877.* In the

---

* "The British Flora Medica. A History of the Medicinal Plants of Great Britain." By Benjamin H. Barton, F.L.S., and Thos. Castle, M.D., F.L.S. A New Edition, Revised, Condensed, and partly Re-written. By John R. Jackson, A.L.S. Chatto and Windus, Piccadilly. 1877.

GATHERING CINCHONA BARK.

forty years which passed between the appearance of the two editions a complete change had taken place in medical and therapeutical science, and the editor of the new edition, in reference to this change, stated his reason for retaining in a condensed form the opinions of ancient writers on the subject-matter under consideration, that they were every year becoming less known, and only to be found in the pages of old and rare books. Thirteen years have now passed since the preface was written to this new edition, and things have probably altered as much in this short period as in the whole forty years previously—new drugs are being dis- covered almost weekly, or some fresh property detected in old sources that was not previously known to exist. By far the largest proportion of new medicinal plants find their way from America, some of which have attained an acknow- ledged pharmaceutical reputation, while many have been found, upon careful trial and experiments in the cure of certain diseases for which they had been recommended, to possess no real or active properties.

Before, however, treating of new drugs in detail, it will be more in place here briefly to review the greatest work of introduction of any useful plant in India and many of our colonies that has been effected in any previous century—we allude, of course, to the introduction and establishment of the cinchonas. The history of this great work has been so often and so fully recorded by the many persons engaged in it that we shall content ourselves with a mere sketch or outline.

The reputation of cinchona bark for the cure of fever was at an early period known to the Spanish Jesuits, and when the Countess of Chinchon, wife of the Viceroy of Peru, fell ill with fever, the bark was administered to her, and speedily effected a cure ; its wonderful properties soon became known, and in 1638 its reputation spread through- out Spain under the name of Jesuits' bark, and for many years the ground bark, mixed with port wine, was a favourite medicine. In the course of time, however, the method of separating the active principles of the bark in the form of alkaloids as quinine, cinchonine, cinchonidine, quinidine, etc., became known, and the first of these—namely, quinine, or sulphate of quinine—soon assumed a most important position as a drug, the demand for which increased so rapidly, not only for consumption in our own country, but also for exportation to India and the colonies, that it realised a very high price, and led to a very great increase in the demand for bark from the South American forests, to meet which trees were cut down in ever-increasing quantities; and as no steps were taken by the governments of the states in which the trees grew to prevent this, or to establish fresh plantations, it was apparent that at no distant date the supply of cinchona-bark must fail, and the most valuable medicine in all the pharmacopœia be entirely lost. Consequently in 1839 the advisability of introducing the trees for cultivation in India was suggested by Dr. Royle; nothing, however, came of this suggestion for at least twenty years. In the meantime the Indian Government were paying enormous sums for quinine—as much as £7,000 in 1852, increasing in 1857 to £12,000. This fact, together with that of the absolute extinction of the plants being in the near future, caused the Indian Government to seriously consider the question, and to adopt arrangements for the introduction of the trees into India. In 1859 Mr. Clements Markham received instructions from the Secretary of State for India to undertake the necessary arrangements for obtaining plants from the South American forests of those species of cinchona which were known to be the most valuable for the production of quinine, and to transmit those plants to India for experimental cultivation. For this purpose two expeditions were organised, one under the direct superintendence of Mr. Markham himself, and the other under the superintendence of Mr. Richard Spruce, an assiduous botanical collector, who had up to that time done much towards the elucidation of Brazilian botany, and who was then residing in New Grenada. In this work, Mr. Robert Cross, at that time employed in the Royal Gardens, Kew, was associated with Mr. Spruce, and to these three gentlemen is due the successful introduction of the cinchona plant into India. The energy and zeal shown by Mr. Cross in this and in subsequent expeditions has been commented upon in the literature of the subject, a considerable share of the success of the whole scheme being due to his knowledge and perseverance. Mr. Cross has since made several expeditions, under orders of the Indian Government, to the different Andean regions to obtain plants of such species of cinchona which are known to be richest in quinine; the plants being brought home and then transmitted to India under Mr. Cross's own personal care, and the results of all his expeditions have been eminently successful.

Referring to the reports on Mr. Markham's expedition, Sir W. J. Hooker in his "Report on the Progress and Condition of the Royal Gardens at Kew during the year 1863" says, "Mr. Markham informs me that in the nurseries on the sites selected by him on the Nilghiri Hills only three years ago, there were on the 1st December last 259,396 plants, of which 66,622 were planted out; that the tallest plant is nearly 10 feet high; that two plants of *C. succirubra* are in full flower; and, further, that 6,562 plants have been distributed to private individuals.

"The bark from some plants has been analysed by J. E. Howard, Esq., and the results have been entirely satisfactory.

"In the Darjeeling plantations, Himálaya, under the superintendence of Dr. Anderson, there are 8,000 plants; and private applications for plants have already been made to that gentleman for the enormous number of 1,500,000."

This indicated not only the great interest taken in the experiment at that early stage, but also proved that the plants in their new home lost little or nothing of their valuable properties, an incentive to private planters to embark in their cultivation. During the following year, 1864, plants were propagated in the Nilghiris at the rate of 30,000 to 40,000 per month; while in Ceylon a stock of 190,000 plants had been obtained, and applications received for 28,500 plants.

In the meantime, plantations had also been established in some of the colonies; and in Jamaica 400 plants had been planted out, and were reported as growing satisfactorily.

In 1865 seeds of *Cinchona officinalis*, which furnishes the pale cinchona or crown bark, were ripened in Ceylon and were distributed to Jamaica, Trinidad, Mauritius, Cape of Good Hope, Queensland, and other places; while in India, in the Nilghiris, at Calcutta, and at Darjeeling the plantations were immensely extended.

In 1867 the plantations in Jamaica contained 30,000 plants; and three years later—namely, in 1870—Sir Joseph Hooker reported as follows :—" The success of the cinchona experiment is now fully established in the Sikkim Himálaya, the Nilghiris, Khasia Mountains (East Bengal), Ceylon, and Jamaica. The bark from the first-named localities has commanded a price equal to the Peruvian in the English market, nineteen cases of red bark from Darjeeling having been bought by Messrs. Howard and Sons for 1s. 9d. per pound, which these gentlemen inform me is what South American bark of the same age would have fetched. No less than a ton of prepared bark has been sent to London from Ceylon, the produce of seeds sent to Dr. Thwaites from Kew in 1861."

In November, 1873, 25,000 pounds of bark from the Nilghiri plantations were sold in London at an average price of 2s. 10d. per pound, the total sum realised being £3,490. One parcel of bark of *C. officinalis* fetched the unusually high price of 5s. 9d. per pound. In the same year 33,000 pounds of dry bark of this species were used on the Nilghiris by Dr. Broughton, the Government quinologist, in the manufacture of a cheap febrifuge, the estimated value of which was £3,300. It was further estimated that 50,000 pounds more bark would be harvested before the end of the year, the value of which would be about £6,700. The total income from the Nilghiri plantations alone for the year 1873 was thus calculated at £13,490, a very good return for the short time the trees had been under cultivation.

In 1877 the first crop of cinchona bark was sent to London from the Jamaica plantations, Mr. Thomson, the superintendent of the Botanical Gardens, remarking at the time that it was abundantly proved that several species of cinchona were eminently fitted for cultivation in Jamaica, so that the enterprise might be considered as having passed from the experimental stage to that of an established agricultural industry. In 1878 a parcel of bark of *Cinchona succirubra*—the red bark—of Jamaica growth was sold in London at 2s. 10d. per pound, being a higher price than that reached by either East Indian or Ceylon bark sold at the same time.

Owing to the success that attended the growth of *Cinchona succirubra* in the Government plantation at Darjeeling, the cheap preparation known as "Cinchona Febrifuge" was begun to be manufactured from it in India in 1877. This preparation was stated to contain "all the febrifugal alkaloids of that species (*C. succirubra*) in the relative proportion in which they exist naturally in the bark."

Dr. King, of the Calcutta Botanic Gardens, writing in 1878, says "Increased experience in the use of the new medicine during the past year has served to establish a large amount of confidence in it. Complaints of its nauseating are now rarely heard, and many medical practitioners of experience affirm that instead of being less potent in the cure of fever than quinine, the cinchona febrifuge is quite equal, if not superior, to the more expensive drug. The proper dose of the febrifuge is, moreover, found to be about the same as, or even less than, the dose of quinine.

" Quinine during the year 1877-78 was very high in price, and in Calcutta it, for some time, stood at 20 rupees per ounce, a sum for which 16 ounces of the febrifuge were always obtainable. The saving to Government by the substitution of cinchona febrifuge for quinine in their hospitals and dispensaries has already been considerable; I calculate that, at a moderate estimate, it amounts to over three lakhs of rupees (£30,000 sterling)."

Again in 1881 Dr. King reports that in the previous year 5,500 pounds of the febrifuge were used in the Government hospitals and dispensaries, in substitution of quinine; and shows by a statement of figures that there was effected, up to that period, a clear saving to the Government of sixteen and a quarter lakhs of rupees.

This saving alone would have sufficiently justified the action of the Government in introducing the plants into India, but the more important one of cheapening an indispensable medicine, and bringing it within the reach of all, must always be borne in mind as the grand result of the cinchona extension.

Up to this period the plants that had been introduced into India and the colonies were mostly the pale cinchona or crown bark (*Cinchona officinalis*), the red bark (*Cinchona succirubra*), and the yellow bark (*Cinchona Calisaya*). In the " Kew Report for 1878," however, the history is given of the introduction of the Columbian barks into India, which is a matter of so much interest that we are compelled to

F

quote it in its entirety. " The Indian Government sent Mr.
R. Cross to New Grenada for the purpose of bringing to
England, for eventual transmission to India, plants of the
species of *Cinchona* yielding the 'Soft Columbian' and

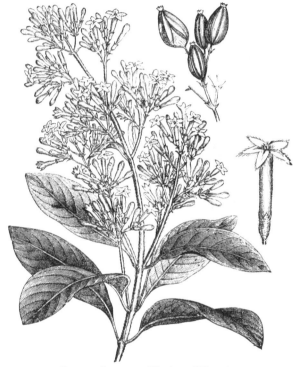

YELLOW CINCHONA (*Cinchona Calisaya*).

'Hard Carthagena' barks of commerce. He arrived in this
country in March of last year with five Wardian cases containing 400 plants of the former and 200 of the latter.
The hard Carthagena included as many as six different

kinds. The barks of all, however, were very carefully
analysed by Mr. Howard, the well-known quinologist.
With regard to the soft Columbian known as ' Calisaya of
Santa Fé,' Mr. Howard reported that the bark analysed, and
which was taken from the rejected cane-like shoots brought
home by Mr. Cross, ' was of the very best description, and
such as indicates the probability of a much larger production
of alkaloid in the bark of more mature and developed trees.'
This bark yielded 6·24 per cent. of alkaloids, of which 3·25
per cent. was quinine, and 1·90 was cinchonidine. Mr.
Howard considers that ' if the young plants can be safely
conveyed to India, and established there, it may not im-
probably prove *second to none.*'

" Of the ' Hard Carthagena ' bark plants, the only one
which Mr. Howard considered worthy of attention was the
kind from Coralis Inza, in the Magdalena valley. This
yielded 4·75 per cent. of alkaloids, of which 1·88 was
quinine and 1·18 was cinchonidine. ' If a free grower, as
I think would be the case, it might be well worth
naturalising in India. The bark has met with a ready
sale in commerce.'

" The plants were placed under Mr. Cross's charge at
Kew, where every facility was afforded him for establishing
and propagating them. On the 16th October of last year
he reported as follows to the Under-Secretary of State :—

" ' On arrival in this country in March, the plants of
the " Calisaya of Santa Fé," carried all the way from the
banks of the Caqueta River, were thought to be in a weak
state. I am glad to state that, although the collection is
now somewhat reduced, there are in all forty plants more
or less growing and rooting, and which I am convinced will
soon become good established plants.

" ' The majority are from cuttings, but there are some
also from original imported root-pieces growing also,
although these in general seemed unwilling to take root

F 2

freely. The dry weather of summer was not so favourable
for the development of growth, but the chief cause was
the diminished vitality of the root-pieces, which were
carried overland so great a distance. When it is considered that these were dug up, and brought from the damp
forest whence rises one of the most important tributaries of
the Amazon, across the Eastern Cordillera down to
Popayan, where, in order to check the growth, they were
alternately covered up and exposed for nearly three
months, then carried down to the hot Cauca valley to the
Pacific, after which there was a month of sea voyage, I
think the result will be deemed rather remarkable. "

At the close of the winter of 1878-9, the stock of
plants remaining at Kew were reduced to fifteen of the
Calisaya of Santa Fé, and ten of the hard Carthagena;
these, however, were considered sufficiently vigorous for
transmission, with proper care, to India, and a portion of
them were accordingly sent during the summer and autumn
of 1879, besides which one plant of the hard Carthagena
bark was sent to Jamaica, which grew and was increased,
so that in 1883 Mr. Morris reported, " We have now a
large number of well-established plants growing at our
highest elevation, and as plants they are the strongest and
finest on the plantations."

The Santa Fé plants having succumbed soon after
leaving Ootacamund, it was decided by the India Office
that Mr. Cross should take out under his own personal
care the three plants that still remained at Kew, together
with a case of Carthagena plants. He left in September,
1880, and subsequently reported from Ootacamund that
the plants were " progressing very satisfactorily, and may
now be regarded as perfectly safe." Subsequent reports,
however, have shown that the plants have not made themselves at home in India.

Under the name of Ledger Bark, the Bark of *Cinchona*

Calisaya var. *Ledgeriana* began to attract a considerable amount of attention about the year 1880. The following history of it is given in the "Kew Report" for that year :—" The seed was collected in June, 1865, by a trusted native servant of Mr. C. Ledger, from fifty trees growing on the almost inaccessible banks of the River Mamore in Bolivia, a place to which no botanist has ever approached more nearly than 100 miles. It was sent to London to the care of Mr. Ledger's brother, who sold half to the Dutch Government, for Java, and half to Mr. Money, a cinchona planter on the Nilghiri hills." From the seeds purchased by the Netherlands Government and sent to Java some 20,000 plants were raised. The name *Ledgeriana* appears to have been a plantation name (first printed about 1873) used to distinguish the progeny of Mr. Ledger's seed ; the plant, however, was described in 1876 by the late Dr. Weddell as *C. Calisaya* var. *Ledgeriana.* Dr. Trimen has since defined it as a distinct species, see *Journal of Botany* for 1881, pp. 321-25.

The purchase of these seeds put new life into the cinchona cultivation in Java. The bark sent thence up to 1872 had been almost entirely of an inferior quality, but in 1873 it was found that the new kind gave 8 per cent. of alkaloids, of which from 5¼ to 6½ was quinine. In the following year this was greatly exceeded, one sample giving a result of 12·97 alkaloids, 11·01 being quinine ; while other samples are reported as yielding the extraordinary amount of 13·7 per cent. of quinine. The tree was consequently cultivated in Java to a very large extent, the plantations containing in 1880 no less than 443,270 trees, and the nurseries 283,650 plants. From the seeds sent to India 60,000 seedlings were raised. Dr. King, reporting about this time to the Bengal Government, says, " After seeing the Java Ledgerianas I have no doubt that our three or four best kinds of Calisaya are precisely the same as some of the forms of Ledgeriana cultivated by the Dutch ; " and he thinks that the smaller yield of quinine from the Indian plants is due to the differences of climate and soil.

From seeds sent to Jamaica a large quantity of plants were raised, and in 1881 were reported as doing well.

Under the name of CHINA CUPREA a new febrifuge bark made its appearance in the English market in 1880. When first introduced it met with no purchasers, and it was even said that a quantity of it was thrown on to the roads in the docks in London to be broken up and destroyed. Shortly after this, however, it began to attract attention in consequence of its high yield of quinine. Mr. R. Thomson, writing from Bogota in December, 1880, says, the plant " was found a few months ago in the state of Santander. It exists in great abundance over several hundred miles of hills, and considerable quantities have been already exported. The bark is peculiar ; it is hard like cinnamon, and contains essential oil in considerable proportion. The quantity of sulphate of quinine averages from 2 to 3 per cent. ; 1,900 men have been collecting this bark, and making roads through the forests in which it abounds. But this will now be checked, inasmuch as the Government has just enforced a tax of 20 dollars on every ' cargo ' to be exported. A cargo weighs about 250 pounds, a mule load. This species is further remarkable from the fact that it grows at the low elevation of from 2,000 to 3,000 feet above the sea—it will therefore become amenable to cultivation in most tropical countries, i.e., those having hills of a moderate elevation. It seems strange that this species should have been left so long undiscovered ; perhaps it is owing to its peculiar character, or perhaps to the fact that the district in which it grows is surrounded and possessed by wild cannibal Indians."

Efforts were made at Kew to obtain seeds of this plant for distribution, and to discover to what species it was to be referred. In 1882 seeds were sent to Ceylon, Calcutta, Fiji, Jamaica, Mauritius, and Seychelles, but few of them, however, germinated. Since then M. Triana has described cuprea bark as being derived from a species of *Remijia*, a genus allied to *Cinchona*. This identification was interesting, as proving that the presence of febrifugal alkaloids is not confined alone to the genus *Cinchona*. Analysis showed that the better sorts of cuprea bark contained on an average 1·8 to 1·9 of sulphate of quinine, very rarely reaching 2 per cent., and in some varieties almost entirely absent. The reason this bark has found favour with manufacturing chemists is because of its freedom from cinchonine, and the readiness with which it pulverises. It is, however, very little used at the present time. *Remijia pedunculata* and *R. Purdieana* are said to furnish cuprea bark.

IPECACUANHA (*Cephaelis Ipecacuanha*).—This plant probably ranks next in importance to the cinchonas, both as regards the medicinal value as well as the interest attached to the attempts to introduce its cultivation in India. True ipecacuanha is a creeping herbaceous plant belonging to the same natural order as cinchona—namely, Rubiaceæ. It is a native of Brazil, and produces long wiry roots marked by annulated rings ; these roots creep for some distance beneath the surface of the ground. For commercial purposes they are dug up, cleansed of the adherent earth, and carefully dried. Ipecacuanha is a valuable expectorant and emetic, and is of very great importance in the treatment of dysentery. It is a singular fact that so long ago as 1648 it was pointed out by Marcgrav and Piso that the powdered root of ipecacuanha was a specific cure for dysentery. This information, however, appears not to have been acted upon till 1813, when it was confirmed by Surgeon Playfair, and again in 1831 a series of reports were published by the Madras Medical

Board showing the effects of ipecacuanha in hourly doses
of five grains, till a hundred grains were often ad-

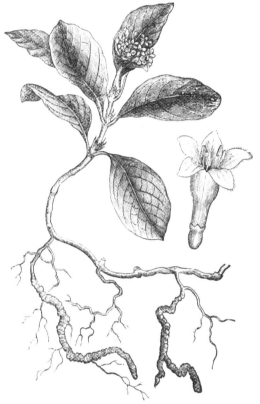

IPECACUANHA (*Cephaelis Ipecacuanha*).

ministered in a very short period. These facts again lay
in abeyance till about twenty-five years ago, when serious
attention was given by the Indian Government to the

value of the plant, and to the necessity of introducing it for extensive cultivation into India. A plant was taken thither from Kew in 1866, which plant, however, died in 1868, fourteen plants having been propagated from it in the meantime. In 1871 five of these plants had been increased to 400 ; and during the year 1870 a large number of plants were transmitted to India from the Edinburgh Botanic Garden, as well as from Kew ; and in 1873 the stock had been increased by the method of leaf propagation in Sikkim Himalaya to over 6,000, at which time it was also being experimented upon in Ceylon. The system of propagation by breaking up the rhizome into small portions, from each of which a plant was grown, resulted in increasing the number in a comparatively short time to about 63,000 plants. Time, however, has proved that while the plant can be propagated to almost any extent, it does better under glass or in frames ; and the slowness of its growth, and the consequently small yield of roots, do not recommend it for profitable culture in India. More recently experiments have been made in Sikkim to cultivate the plant in shady places, under trees. In 1876 a quantity of the root was dried and prepared by Dr. King for use in the Medical College Hospital, Calcutta, with the result that its efficacy was proved to be quite as great as the best Brazilian ipecacuanha.

The cultivation of ipecacuanha has also been attempted in Burmah and Singapore ; and the report of its progress in the Straits Settlements in 1886 was that it grew there " with all the luxuriance of its native country when a proper situation is hit upon," and under these conditions it forms a compact little shrub about eighteen inches high. Plantations containing thousands of plants in excellent health are reported from Johore, and the produce of these plantations has found its way into the London market.

Ipecacuanha is imported into this country, chiefly from Brazil, to the extent of about 65,000 pounds annually, representing a money value of about £15,000.

JALAP (*Ipomœa purga*).—Though this is by no means a new introduction, the purgative properties of jalap-root having been known in Europe since the sixteenth century, the introduction and cultivation of the plant in the East and West Indies claim a note in this place, if only to put it on record. The plant belongs to the natural order Convolvulaceæ, is of a climbing habit, and is a native of the eastern declivities of the Mexican Andes. So far back as 1870 some tubers were planted at Ootacamund, which grew remarkably well, indicating that, if found necessary, the plant could be extensively cultivated in that part of India. The propagation of the jalap plants in the Ootacamund Garden from root and stem cuttings was considerably extended, till in 1877 there were upwards of 25,000 plants permanently planted out, covering an area of about five acres. The experience gained in jalap cultivation in Madras showed that the plants preferred a tolerably rich, dry, and friable loamy soil. Well-drained grass land is best, and it should be laid out in terraces ten feet wide ; the ground should be dug over to the depth of two feet, and left exposed to the action of the sun for two months. After being drilled and manured, it should be planted with potatoes ; and when these are lifted, the jalap tubers are planted in rows on ridges a few inches high to prevent water from becoming stagnant around them ; when the plants have become established, ordinary garden culture is all that is required.

The plants being of a climbing habit require stakes, or trellis, or some such support as is given to peas. The stems die down annually, and the tubers remain dormant for two or three months. The plants give off a number of underground shoots, upon which tubers are formed, and from which fresh plants are readily raised. One acre of land planted as described would yield at the expiration of three years 5,000 pounds of green tubers, which, when thoroughly dried, would give 1,000 pounds of jalap powder. In Jamaica a

RHUBARB.

similarly successful trial was made in 1881 in the culture of jalap ; five acres of land were also planted in this island, and the crop yielded 3,000 pounds of green tubers, resulting in 1,077 pounds of cured jalap. Care in drying or curing the roots seems to be as necessary as good culture of the plants. In the report of the Jamaica experiment it is said that "the application of too much heat was apt to change 'the character of the starch grains and convert them into an amorphous mass.' The small tubers were dried whole, while the larger ones were sliced or gashed." These latter were not approved of in the London market, while in New York they fetched higher prices than with us, and no objection was made to the slicing.

Jalap is imported chiefly from Vera Cruz, and is brought into the United Kingdom to the extent annually of 180,000 pounds.

RHUBARB (*Rheum officinale*).—This fine species was discovered in South-eastern Thibet, where it is said to be often cultivated for the sake of its root for use in medicine. The plant was first obtained about 1867. It was first grown in this country in 1873 by the eminent pharmacologist, Daniel Hanbury; after which it was cultivated for medicinal purposes at Bodicote near Banbury, where it is still grown. At one time it was supposed to be the source of the best quality of medicinal rhubarb; more recent information, however, has shown that this is the produce of *Rheum palmatum*, a large perennial herb first found wild by Colonel Prejavalsky in 1872-3 in the Tangut district of Kansu. It is said to extend over a large tract of country, but not to be cultivated, though sometimes grown as a garden plant. It is this species that is now known as the real source of the Russian or Turkey rhubarb of commerce, and is derived alone from the provinces of Shensi, Kansu, and Szechuen. So great is the demand for medicinal rhubarb at the present time that as many as 625,000 pounds are annually sent from Hankow to Shanghai, 350,000 pounds of which find their way into the United Kingdom.

SOCOTRA ALOES (*Aloe Perryi*).—The discovery of the source of true Socotrine aloes is one of considerable interest

and importance, and may be briefly stated thus :—The substance called aloes was known to the Greeks so far back as
the fourth century before the Christian era as a product of
Socotra, and its source has been attributed by all writers,
even the best authorities, to *Aloe Socotrina*, a species
cultivated in England for more than a hundred years. It
was not, however, till 1878 that the first authentic specimen
of Socotra aloes was brought to this country by Mr.
Wykeham Perry. The plant reached Kew in a living state,
but did not survive. Professor Bayley Balfour, however,
during his visit to Socotra in 1880 obtained a supply of both
living and dried plants, as well as of the aloes itself in all
stages of preparation ; these specimens are now contained in
the Kew Museum, and one of the living plants flowered in
the succulent house at Kew in 1881, and has been named
by Mr. J. G. Baker *Aloe Perryi* after its discoverer.

DRAGON'S BLOOD (*Dracœna schizantha* and *D. Ombet*).—
Dragon's blood of Africa, though known in medicine from
the earliest historical times, now rarely finds its way into
commerce, and until recently little or nothing has been
known of its history. In 1871, however, Mr. Baker
described a species of *Dracœna* found in Somaliland, and
yielding dragon's blood, under the name of *D. schizantha*,
and he also identified a species of *Dracœna* found in Socotra
as *D. Ombet ;* this tree is described as growing at an elevation of 1,500 feet above the sea ; it attains a height of
twenty feet, with spreading drooping branches of a mushroom
form. The resin is obtained by scraping the bark, and after
fifteen or twenty days it exudes. It is exported from Aden
chiefly to Bombay, where it is used by goldsmiths.

NEPAL and BENGAL CARDAMOMS (*Amomum subulatum*).
—Though it is well known that the bulk of the cardamoms
of commerce—namely, those grown in Southern India—are
the produce of *Elettaria cardamomum*, the sources of some
of the other kinds are by no means definitely settled. Those

known as Nepal and Bengal cardamoms, the latter of which were supposed to be the produce of *Amomum aromaticum*, were proved, together with those of Nepal, by Dr. King, of the Calcutta Botanic Gardens, in 1877, to be obtained from *Amomum subulatum*. It is interesting also to learn from the same authority that the fruits of the true *Amomum aromaticum* of Roxburgh, which were doubtless in the Indian markets in his days, are now unknown in commerce.

BLACK SNAKE ROOT, or BLACK COHOSH (*Cimicifuga racemosa*).—A perennial herb common in the woods of Canada and the United States. The roots were first introduced into medical practice in America in 1823, and into England about 1860. It is administered in the form of a tincture in rheumatic affections, and is used also in dropsy, phthisis, and in chronic bronchial affections.

RHATANY ROOT (*Krameria triandra*). — A woody branched shrub about a foot high, belonging to the natural order Polygaleæ, native of the sandy declivities of the Bolivian and Peruvian Cordilleras. The root is gathered chiefly to the north, north-east, and east of Lima, and also in the northern part of Peru. Rhatany root is a powerful astringent, but is not much used at the present time in this country. It first appeared at a London drug sale at the commencement of this century, and formed part of the cargo of a Spanish prize. It was first described in the *Medical and Chirurgical Review* in 1806.

BUCHU LEAVES (*Barosma crenulata, B. serratifolia*, and *B. betulina*).—These are all shrubby plants, natives of the Cape of Good Hope. Buchu leaves are chiefly administered in diseases of the urino-genital organs ; and were introduced to the medical profession of London in 1821. The plants belong to the natural order Rutaceæ.

BAEL FRUIT (*Ægle Marmelos*).—A widely spread Indian tree of moderate height It is held sacred by the Hindoos, and is often planted in the neighbourhood of temples. It

CARDAMOM (*Elettaria cardamomum*).

belongs to the natural order Rutaceæ, and is a close ally to the orange, which fruit it much resembles. Though it has long been known in India as a remedy for dysentery and diarrhœa, it was not till about the year 1850 that it began to attract attention as a medicine in Europe.

QUASSIA WOOD (*Picræna excelsa*).—A tree of 50 or 60 feet high, belonging to the natural order Simarubeæ. It is common in Jamaica, and is also found in Antigua and St. Vincent's. Quassia wood was originally derived from *Quassia amara*, a tree botanically allied to that under dis-

QUASSIA (*Picræna excelsa*).

cussion, and a native of Panama, Venezuela, Guiana, and Northern Brazil ; but in 1809 it was superseded by the wood of *Picræna excelsa*, which is a tree of much larger size, and more abundant. The wood is used as a stomachic and tonic, and is usually seen in the druggists' in the form of chips or turnings. The bitter cups, which were common a few years ago, and which gave a bitter draught by allowing water to stand in them for a short time, were turned out of this wood. Quassia has a slightly narcotic effect on the higher animals, but is poisonous to flies.

ROHUN BARK (*Soymida febrifuga*).—A large tree of Central and Southern India, belonging to the natural order Meliaceæ. The bark is used in India as an astringent tonic and antiperiodic, in intermittent fevers, general debility, diarrhœa, and in the advanced stage of dysentery. It was sent by Roxburgh to Edinburgh at the end of the last century, for trial, and was introduced into the Edinburgh Pharmacopœia in 1803, and into the Dublin Pharmacopœia in 1807.

COWHAGE or COW-ITCH (*Mucuna pruriens*).—A strong climbing leguminous plant, common throughout the tropics of India, Africa, and America. It produces a large number of pods from 2 to 4 inches long and about half an inch wide. They are slightly curved, of a dark brownish colour, and thickly covered with stiff sharp hairs, which are easily detached from the valves, and penetrate the skin, causing an intolerable itching. These hairs have long been known as a vermifuge ; and in this country began to attract attention at the latter part of the last century. As a drug, cowhage was introduced into the Edinburgh Pharmacopœia in 1783 and into the London Pharmacopœia in 1809. It is now seldom used in European practice.

WILD BLACK CHERRY BARK (*Prunus serotina*).—A plant of variable habit, widely spread over North America, forming a shrub in some localities, and in more favourable situations growing to a height of 60 feet. It belongs to the natural order Rosaceæ. The bark has a high reputation in America as a mild tonic and sedative, and was introduced to notice in this country in 1863, but is not much used with us in medical practice.

CHERRY LAUREL LEAVES (*Prunus Lauro-cerasus.*)—This well-known evergreen shrub thrives well with us, and in other countries where the winters are not severe. It is a native of the Caucasus provinces of Russia, North-western Asia Minor, and Northern Persia, and has been introduced

G

on account of its ornamental appearance to all the more temperate parts of Europe. The leaves, cut up and distilled with water, yield a volatile oil like bitter almond oil, which contains hydrocyanic acid. They are also used for making cherry laurel water, and were introduced to the British Pharmacopœia for this purpose in 1839.

CAJUPUT OIL (*Melaleuca Leucadendron*, var. *minor*).— This is a large myrtaceous tree, abundant and widely spread in the Indian Archipelago and Malay Peninsula. The oil, which is obtained from the leaves by distillation, is chiefly prepared in the island of Bouro, one of the Moluccas. It first made its appearance at Amsterdam about 1727, was admitted to the Edinburgh Pharmacopœia in 1788, but does not appear to have become an article of commerce with us until 1813. It is used externally as a rubefacient and occasionally given internally as a stimulant and dia-phoretic.

GAMBIER or TERRA JAPONICA (*Uncaria Gambier*).— The plant yielding this substance is a strong-growing climber, belonging to the natural order Rubiaceæ, and native of the countries bordering on the Straits of Malacca. It is also grown in Ceylon. For commercial purposes plan-tations were formed for its cultivation in Singapore so far back as 1819, and at the present time it is grown there on a very large scale. Gambier is prepared by boiling the leaves and young shoots in water in an iron pan, after which the decoction is evaporated to the consistence of a thin syrup, when it is poured into buckets and submitted to a kind of churning action, when it becomes thick, and sets into a mass resembling a soft yellowish clay, which is put into square boxes and cut into cubes, and dried, when it is ready for ex-portation. It was first brought to notice in this country about the year 1807, and is used medicinally as an astringent. It is also largely used in dyeing and tanning.

INDIAN TOBACCO (*Lobelia inflata*).—An erect annual or

biennial herb, 9 to 18 inches high, widely distributed over the Northern United States. The plant belongs to the natural order Campanulaceæ. The dried herb is imported into this country in pieces of varying sizes, and compressed into oblong packages. In moderate doses it is a powerful emetic, but in large doses it acts as an acro-narcotic poison. It is administered in spasmodic asthma. Its properties have long been known in America, but it was not till about 1829 that it was introduced to England.

CHIRETTA (*Swertia chirata*).—An annual herb belonging to the natural order Gentianeæ, and native of the mountainous regions of Northern India. The whole plant possesses a strong bitter taste, and has long been held in high repute by the Hindoos as a tonic. About 1829 it began to attract some attention in England, and was admitted to the Edinburgh Pharmacopœia in 1839. It is a pure bitter tonic, without aroma or astringency, and is used in this country chiefly in the form of tincture. It is also said to be used, in the place of gentian, to give flavour to the compound cattle foods now so general.

BELLADONNA or DEADLY NIGHTSHADE (*Atropa Belladonna*).—This well-known herbaceous plant is very widely spread, not only in this country, but also through Central and Southern Europe, Caucasia, and Northern Asia Minor. The roots are chiefly used for the preparation of atropine, employed in ophthalmia for dilating the pupil of the eye, and for making a liniment for neuralgic pains ; for this purpose it was introduced about 1860. The leaves were introduced into the London Pharmacopœia in 1809, for the preparation of extracts and tincture.

BEBEERU or GREENHEART BARK (*Nectandra Rodiœi.*)— A large hard-wooded forest tree of British Guiana, belonging to the natural order Laurineæ. The thick bark contains an alkaloid known as *Beberine*, and has been recommended as a bitter tonic and febrifuge. It first attracted

G 2

attention about 1835, and the alkaloid was further examined in 1843. The supply of Greenheart bark to the English market is very irregular.

MATICO (*Piper angustifolium*).— This is a shrub belonging to the natural order Piperaceæ, native of Bolivia, Peru,

BELLADONNA (*Atropa Belladonna*).

Brazil, Venezuela, and New Granada. Matico, as seen in commerce, consists of the broken and compressed leaves, which are very thick and very rough on the surface ; they have a pleasant, somewhat pungent odour, and a bitterish aromatic taste. They are used, either softened in water or reduced to a powder, to stop bleeding, and an infusion prepared from them is also administered for internal hæmorrhage. They come by way of Panama in bales or serons. Matico was first brought to notice in this country by a Liverpool physician in 1839.

Though the source of Matico is generally believed to be the plant mentioned above, the leaves of other allied species no doubt are often mixed with them. Thus, at the close of the year 1888, a consignment of Matico leaves reached the London market which proved to be derived from *Piper Mandoni*.

LARCH BARK (*Larix europea*).—The bark of this well-known coniferous tree, which has been known for a very long time to possess astringent properties, and is in consequence used for tanning, was first brought to notice in this country in 1858 as a stimulating astringent and expectorant. It is used chiefly in the form of a tincture.

ARECA or BETEL NUTS (*Areca Catechu*).—This is a palm growing to a height of 40 or 50 feet, with a straight smooth trunk from 1 to 2 feet in circumference. The tree is probably a native of the Malayan Archipelago, where it is also cultivated, as well as in the warmer parts of the Indian Peninsula, Ceylon, and the Philippine Islands. The seeds of this palm, which are known as Areca nuts, are about the size and appearance of a small nutmeg somewhat flattened at the base, and like the nutmeg they are ruminated or marked throughout their substance by dark irregular lines. They possess astringent properties, and are held in high repute among Asiatics as a masticatory, as well as for strengthening the gums and sweetening the breath. It has attracted some attention of late years as a tænifuge for the expulsion of tapeworm, given in doses of from four to six drachms in milk, and has been used in this country for this purpose since 1867.

INDIAN POKE-ROOT (*Veratrum viride*).—A plant belonging to the natural order Liliaceæ, and common in swamps and low grounds from Canada to Georgia. The purgative and antiscorbutic properties of the plant have long been known in North America, and in 1862 the roots, or more properly the rhizomes, were introduced into this country as a cardiac, arterial, and nervous sedative.

COLCHICUM SEEDS (*Colchicum autumnale*). — A well-known liliaceous plant in meadows and pastures in this country, as well as over a large portion of Middle and Southern Europe. The corms are the source of the specific known as wine of Colchicum, and have been used in medicine from early times. In 1820 the seeds were introduced into medical practice on account of their being said to have a more certain action than the corm, and were introduced into the Pharmacopœia in 1824.

CHAPTER VI.

NEW DRUGS.

To give a complete list of the new remedies that have been brought to the notice of the British pharmacist during a comparatively recent period would occupy much more space than would be justifiable, for scarcely a week now passes without the appearance of a note on some novelty in the pages of the medical and pharmaceutical journals. It will therefore suffice to enumerate only those to which most attention has been given, such as those which have already come into use, or which promise to become established medicines. Those which are enumerated below are classified in alphabetical order of their scientific nomenclature.

Abrus precatorius.—A common tropical plant belonging to the natural order Leguminosæ, well known for its small globose scarlet and black seeds, which are used almost everywhere in the tropics for making necklaces, bracelets, and other ornaments, as well as for weights by the diamond merchants in India. These seeds began to attract attention in 1882, having been experimented with on the Continent in the treatment of ophthalmic diseases under the name of JEQUIRITY. In Egypt they are occasionally used as an article of food, and are harmless, but powdered and introduced beneath the skin they rapidly produce fatal effects. The poisonous action is due to the presence of *abrine,* which is rendered inert by heat, and is closely allied to albumin in composition. It is obtainable also from the roots and stem. This plant has recently become known as the weather plant.

Alstonia scholaris.—A tree 50 to 80 feet high, widely diffused in India, Africa, and Australia, and belonging to the natural order Apocynaceæ. The bark is powerfully bitter, and is used by the natives of India in bowel complaints. Under the name of DITA bark it began to attract attention in this country in 1875 as a most valuable antiperiodic and tonic.

An allied species, *A. constricta*, a native of Queensland and New South Wales, and known as the QUEENSLAND FEVER BARK, where it has had a reputation for some time, has also been introduced since 1878, and used as a tonic and febrifuge.

Andira araroba.—Under the name of GOA POWDER a substance was introduced in 1874 to the notice of pharmacists as a cure for ringworm and other skin diseases. The drug was imported into the London and Liverpool markets from Bahia, and consisted of lumps of a yellowish substance, composed partly of powder and partly of pieces of wood. For some time its botanical source remained unknown ; but in 1879 Dr. I. M. de Aguiar published at Bahia a botanical description of the plant under the above name. The active principle of the drug, called Chrysophanic acid, soon obtained for it a reputation in the cure of the diseases referred to, and the drug is still included in the chemists' trade lists.

Aspidosperma Quebracho-blanco.—A tree, native of the Argentine Republic, and belonging, like the last, to the natural order Apocynaceæ, furnishes the Quebracho-blanco or White Quebracho bark of commerce. It is used in various forms of dyspepsia, bronchitis, phthisis, etc., and was introduced to the notice of English pharmacists in 1879.

Cannabis indica.—The common HEMP is well known to be valuable for two distinct economic uses—namely, when grown in cool countries it is valued for its fibre, and when grown in hot countries, for the resin which is secreted all over the plant. In India and other tropical countries this is much used under the names of Bhang, consisting of the dried leaves and slender stalks; Ganja, the flowering

PAPAW (*Carica papaya*).

or fruiting shoots; and Churrus, the resin itself. The

introduction of the Indian drug into European practice is chiefly due to experiments made in Calcutta by Dr. O'Shaughnessy in 1838–39.

Carica papaya.—The PAPAW tree has always had a peculiar interest attached to it in consequence of the statements of travellers that it possessed the extraordinary property of rendering tough flesh tender by merely hanging the freshly-killed meat amongst the foliage of the tree. In the "Natural History of Jamaica" Browne says that meat is quickly made tender by washing it with water mixed with Papaw juice; and if left in the water for ten minutes, the meat will fall to pieces or divide into shreds during the process of cooking. Nothing like real attention was given to this important property till about 1878, since which time it has received considerable notice at the hands of chemists and the medical profession, not only in this country, but in Europe generally, in the treatment of dyspepsia, diphtheria, etc. The native country of the plant is supposed to be the warm part of the American continent, but it is now widely scattered in tropical countries in both hemispheres. The fresh fruits are generally cooked and eaten as a green vegetable in the countries where the plant grows.

Cinnamodendron corticosum.—Under the names of RED CANELLA, MOUNTAIN CINNAMON, or FALSE WINTER'S BARK the bark of this tree has been long known for its stimulant, tonic, aromatic, and antiscorbutic properties. It is a small tree, 10 to 15 feet high, but sometimes growing to a height of 90 feet. It is confined to Jamaica; and though the bark has been well known for so long, the plant remained undescribed till about 27 years ago. Plants have been in cultivation in the Royal Gardens, Kew, and in the Gardens of the Royal Botanical Society, Regent's Park, for some years, and flowered for the first time at Regent's Park in 1874.

Cola acuminata.—This tree, which grows to a height of about 40 feet, is a native of the west coast of Africa,

between Sierra Leone and the Congo, and belongs to the
natural order Sterculiaceæ. The seeds, several of which
are contained in a fleshy fruit four to six inches long, are
the well-known KOLA-NUTS of West Africa, where they are
extensively used for satisfying the cravings of hunger and
enabling those who eat them to endure prolonged labour
without fatigue. Powdered, they are used to clarify stagnant water, which is said to be thus rendered agreeable
to the taste. The trade in Kola-nuts is one of great importance in the Gambia and at Sierra Leone, and of late
years it has spread to Central Africa, and also to the
African shores of the Mediterranean, as well as in the
West Indies. In view of its probably becoming an important plant for cultivation in tropical countries, a number
of plants were propagated at Kew in 1880, and distributed
to Calcutta, Ceylon, Demerara, Dominica, Sydney, Mauritius, Zanzibar, Java, Singapore, and Toronto ; so that
at the present time the nuts are produced in other countries
than Western Africa. Indeed, five or six years ago it was
reported from Jamaica that if a demand should arise for
them in this country, they could be shipped thence to the
extent of several tons a year. The suitability of the West
Indies for the cultivation of Kola-nuts was well exemplified
by the exhibits of fruits and seeds in the Colonial and
Indian Exhibition of 1886, which were very fine, especially
those grown in Grenada. Notwithstanding, however, that
Kola has occupied the attention of English and Continental
chemists and pharmacists for several years past, and important properties and uses have been assigned to it—notably
that of restoring the nerves after a too free use of stimu-
lants, and as an ingredient in the preparation of cocoa
and chocolate, by which the strengthening power of those
beverages is said to be considerably increased, so that a
workman can, on a single cup taken at breakfast-time, go on
with his work through the day without feeling fatigued—

no great demand has up to the present time arisen for it. The Kola-nut plant has been in cultivation in this country for some years previous to 1868, in which year it flowered for the first time at Kew.

Colubrina reclinata.—The bark of this plant, under the name of MABEE BARK, began to attract some notice in this country as a medicine about 1885, in consequence of its being largely used in the West Indies in the preparation of a stomachic drink. The plant is a native of South America.

Copernicia cerifera.—This is the WAX PALM or CARNAUBA of Brazil, the roots of which are said to have diuretic properties, administered in the form of infusion, decoction, or fluid extract. The infusion has an agreeable and slightly bitter taste, and an odour somewhat resembling that of Sarsaparilla. It was introduced to notice in 1875, but is now but little heard of.

Cybistax antisyphilitica.—The leaves of this plant, under the name of CAROBA, are used in Brazil as one of the best alterative, diuretic, sudorific, and tonic medicines. Attention was first directed to it in this country about 1875. From the accounts which accompanied its introduction it would appear to be extremely useful in all kinds of syphilitic affections.

Duboisia Hopwoodi.—The broken leaves of this plant, known as PITURI, have been used by the aborigines of Central Australia from an early period as a stimulating tonic, being chewed by them to strengthen themselves on long journeys or to increase their courage in battle. It was introduced to the notice of the medical world in 1873 as a narcotic stimulant.

Eucalyptus globulus, BLUE GUM of Tasmania, where, as also in Victoria, it grows over 300 feet high. Introduced in 1856, it has become very common in many parts of England, though, with the exception of Cornwall and the

West of Ireland, it cannot stand the winter without shelter.
In the South of Europe it has become familiar, and has been
largely planted in malarial districts in Italy, as, on account
of its rapid growth and antiseptic exhalations from
the leaves, it is said to absorb the moisture from the
ground and purify the air. The leaves have a bitterish,
pungent, camphoraceous taste and smell, due to the presence
of a volatile oil. They have been recommended as a remedy

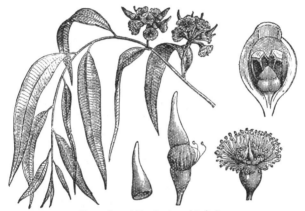

BLUE GUM (*Eucalyptus globulus*).

in fevers. The oil distilled from them is tonic, stimulant,
and antiseptic. It has been used externally as a rubefacient,
also in perfumery for scenting soaps, and internally in
bronchial and diphtheritic affections, under the name of
Eucalyptol. The resin of this species and that of *Eucalyptus amygdalina* forms Australian Kino.

Euphorbia Drummondii.—A prostrate or diffused much-
branched plant of Australia. An alkaloid contained in this
plant, called *Drumine*, has been discovered and applied within
the last year or two as a local anæsthetic.

GATHERING COCA.

Euphorbia pilulifera.—A native of Australia, where it attracted much attention some few years since as a cure for asthma and bronchial affections. The plant being poisonous, care has to be exercised in its proper administration; therefore too strong a decoction must not be used. It was introduced to notice in this country in 1882.

Erythroxylon Coca.—Under the name COCA the leaves of this plant have long been known and used by the Indians of Peru and Bolivia as a valuable nervous stimulant, as well as to prevent hunger, and to enable those who use them to endure long periods of labour without fatigue. These properties began to be noticed in this country in 1874, since which time the leaves have been experimented upon in various ways, till coca has become an important medicine, and both the leaves and preparations from them are now fully-recognised articles of trade, and *Cocaine*, the active principle of the plant, has become an important medicine. The cultivation of the plant has been extended into several of the British colonies, including Jamaica, Trinidad, Zanzibar, Brisbane, etc.

Fabiana imbricata.—This is one of the latest introductions, having been first brought to notice in America in 1885 and in this country in 1886. It is known as PICHE, and is a native of Chili, Peru, and the Argentine Republic, and belongs to the natural order Solanaceæ. It is recommended in lumbago, sciatica, rheumatic neuralgia, irritability of the bladder, etc.

Ferula (Euryangium) Sumbul, an herbaceous perennial, dying after flowering, belonging to the natural order Umbelliferæ. The root, which has a strong musky odour, is known as SUMBUL or MUSK ROOT. The early history of Sumbul cannot be traced; all that is known about it is that it was first introduced into Russia about 1835 as a substitute for musk, and was then recommended as a remedy for cholera. It began to be known in Germany in

1840, and in England in 1850, and was introduced into the British Pharmacopœia in 1867. Its source was unknown till 1869, when the plant was discovered in the mountains south-east of Samarkand, separating Russian Turkestan from Bucharia, at an altitude of from 3,000 to 4,000 feet. In 1872 a root was received at Kew from the Moscow Botanic Garden, which continued to throw up leaves each year till 1875, when it produced a fine flowering stem, but did not ripen fruit, and afterwards died.

Pereira mentions two kinds of Sumbul known to commerce in his time—one called Russian and the other Indian Sumbul, the latter coming to England *viâ* Bombay. All the Sumbul which reaches Europe and the United States at the present time is imported from Russia. As seen in commerce, Sumbul root occurs in roundish pieces or transverse sections; it has a powerful but agreeable musky odour, which is very persistent. It is used in medicine as a nervine stimulant and antispasmodic, and is given in chronic pulmonary affections, hysteria, etc.

Franciscea uniflora.—A shrub belonging to the natural order Scrophularineæ, native of Brazil and other parts of equatorial America. The roots, under the name of MANACA, are used in the preparation of a powerful antisyphilitic, purgative, and diuretic medicine. It was introduced to notice in this country from North America in 1883.

Frankenia grandifolia.—An herbaceous plant of California. Under the name of YERBA REUMA it was introduced in 1879 as a remedy in catarrh, mucous discharges, and in ophthalmia.

Gouania domingensis.—A climbing West Indian shrub belonging to the natural order Rhamnaceæ. It has long been known as CHEW STICK, and used when pulverised as an ingredient in tooth powder. Pieces of the stem, with one end beaten into fibre, have also been used as tooth-brushes. These stems appear to contain saponine. In the

West Indies the whole plant is considered a good antiseptic; a decoction of the root has been used in dropsy. It
was introduced to notice in this country about 1884 for use
in the preparation of an astringent gargle.

Gynocardia odorata.—A large tree of India bearing a
globular fruit about the size of a large orange, and containing numerous seeds, the oil of which is expressed and known
as CHAULMUGRA oil. This oil has been used in India for a
very long time in skin diseases, affections of the joints, etc.
It was not, however, till 1878 that it began to attract
much attention in England, when experiments were made
in many of the London and provincial hospitals, as well as
in private practice, to test its efficacy in rheumatic affections,
skin diseases, consumption, syphilitic affections, etc.; it was
used both externally and internally, the latter in the form
of capsules. A certain amount of success seems to have
attended its use, but of late years it has ceased to attract so
much notice.

Hagenia abyssinica.—A handsome rosaceous tree 50 or
60 feet high, found over the whole table-land of Abyssinia.
Under the name of KOUSSO or KOSSO the flowers have a
reputation as an anthelmintic. Notices first appeared as to
their medical properties in English periodicals during the
years 1839 to 1841, but no supply of the flowers reached
Europe till 1850, when a quantity was brought to London
and offered for sale at 35s. per ounce. Large quantities
were afterwards imported and sold at from 3s. to 4s. per
pound. It was not till 1864 that Kousso was introduced
into the British Pharmacopœia.

Hemidesmus indicus.—A twining shrub of the Asclepiadeæ, native of India and Ceylon. The roots are known
as INDIAN SARSAPARILLA, and have been used for a long
period in native medicine in India. They are said to
have alterative, tonic, diuretic, and diaphoretic properties, and were introduced into the British Pharmacopœia

H

in 1864. They are, however, very rarely employed in this country.

Joannesia princeps.—Under the name of ANDA-ASSU this tree was first brought to notice in 1881 as yielding seeds valuable in Brazil as a purgative and for affections of the liver, jaundice, dropsy, etc. It belongs to the natural order Euphorbiaceæ.

Mallotus philippinensis.—A large shrub or small tree 20 to 30 feet high, belonging to the same natural order as the last, and very widely distributed, being found in Abyssinia, Southern Arabia, throughout India, in Ceylon, Malay Archipelago, Philippines to Australia. The red glandular powder obtained from the fruits is known as KAMALA. It is used as a vermifuge, or rather as a tænifuge in the cure of tapeworm in India, as well as for dyeing silk red. It was introduced into the British Pharmacopœia in 1864.

Marsdenia cundurango.—The bark of this plant, under the name of CUNDURANGO, began to attract considerable attention in America as a remedy for the cure of cancer in 1871. Samples having been sent from Ecuador, its reputation soon reached this country, but it was not till the following year (1872) that its botanical origin became known, when it was described by M. Triana under the name of *Gonolobus cundurango*, which has since been referred to *Marsdenia*, natural order Asclepiadeæ.

For some time Cundurango bark was submitted to numerous experiments, with the result that it was generally pronounced to be of little or no use medicinally in cancer cases. Some interest, however, attaches to it in consequence of its being included amongst the plants used by the natives for the cure of snake-bites, under the name of Guaco. The word Cundurango means "vine of the Condor"—from a tradition of the country that when the condor is bitten by a poisonous snake, it swallows the leaves of this plant and experiences no harm.

Mentha arvensis, var. *piperascens.*—A Chinese herb belonging to the Labiateæ. It yields an oil which contains a large quantity of a crystalline substance known as MENTHOL or PEPPERMINT CAMPHOR. This substance began to attract attention in 1879, since which time Menthol has become an increasing article of trade, and is much used in cases of neuralgia, toothache, etc., by rubbing it on the parts affected. A similar crystalline principle is obtained in India from the oil expressed from the seeds of *Carum copticum.* The Chinese peppermint plant has been recommended for cultivation in England, and especially in Ireland, where the climate is moist and labour cheap.

Myrtus cheken.—An evergreen climber belonging to the natural order Myrtaceæ, and native of Chili, where it is known as CHEQUEN, and is in great repute as a medicine in inflammation of the eyes, in diarrhœa, and other disorders, for which purposes it was introduced into this country in 1881. Though the plant has been cultivated in our greenhouses for many years, it flowered for the first time at Kew in 1866.

Paullinia sorbilis.—A woody climber belonging to the natural order Sapindaceæ, and native of the Northern and Western parts of Brazil. The seeds, which are like small horse-chestnuts, are used in Brazil in the preparation of a beverage and as a medicine. To prepare them the seeds are dried, powdered, mixed with water, and kneaded into a kind of dough, then made into rolls, or moulded into various forms, and known as GUARANA, GUARANA BREAD, or BRAZILIAN COCOA. It is regarded as a tonic, febrifuge, nutritive, and to some extent narcotic. As a nervous stimulant, it is analogous to tea and coffee, and has been recommended in this country in nervous headache, neuralgia, paralysis, and diarrhœa. It can be administered either in the form of a substance, as a beverage, or mixed with cocoa or chocolate. It was introduced to notice in this country firstly in 1856, and again in 1870.

H 2

Peumus Boldus.—A shrub 10 to 20 feet high, native of Chili, and cultivated in gardens in its native country for the sake of its fragrant flowers and leaves. The plant flowers in its native home in autumn, but under cultivation at Kew and the Royal Botanical Society's Gardens, Regent's Park, the flowers have appeared in winter. The plant belongs to the natural order Monimiaceæ, and the leaves, under the name of BOLDO, were introduced to this country in 1874, as an aid to digestion as well as in diseases of the liver. The properties of the plant are said to have been discovered by noticing the beneficial effects upon a flock of sheep that were suffering from liver disease. Having been shut up in a fold which had been recently repaired with the twigs of the Boldo plant, the sheep ate of the leaves and shoots, and recovered very speedily. The leaves, dried and pulverised, are used in Brazil as a sternutory.

Physostigma venenosum.—A perennial climbing plant with a woody stem 50 or more feet high, belonging to the natural order Leguminosæ, and found near the mouths of the Niger and Old Calabar River. The seeds are known under the names of ORDEAL BEANS OF OLD CALABAR, or CALABAR BEANS, and they were first brought to notice in England about the year 1840 by Dr. W. F. Daniell, who in 1846 brought them more prominently forward in a paper read before the Ethnological Society. The poisonous effects of the beans on the human system were noticed by Christison in 1855, and again by Sharpey in 1858. In 1859 a plant was sent by an African missionary to Professor Balfour, of Edinburgh, who described it under the name it now bears. It was not till about 1863 that Professor Fraser discovered that an alcoholic extract of the seed possessed the power of contracting the pupil of the eye, since which time it has been used in ophthalmic practice, as well as in tetanus, rheumatic, neuralgic, and similar affections. The plants are somewhat rare in Africa, being destroyed by order of the Government, except so many as are required to supply seeds for use as an ordeal. They find their way, however, to this country in small quantities from West Africa.

Picramnia antidesma.—Under the name of CASCARA AMARGA the bark of this Mexican tree, which belongs to the natural order Simarubeæ, was first brought to notice in America in 1885, and soon after reached this country. It is said to be useful in syphilis, and as an external application in the treatment of erysipelas.

Pilocarpus pennatifolius.—This is a shrub four or five feet high, belonging to the order Rutaceæ, native of Brazil, and was first found in the southern provinces of Mato Grosso and São Paulo, from whence it was introduced into Europe in 1874, and is now found cultivated in the English and Continental botanical gardens. Under the name of JABORANDI a new drug was introduced to the notice of British pharmacists in 1874. Jaborandi, however, appears to be a comprehensive name in South America, and is applied to a number of widely different plants. The determination of the source of the ordinary Jaborandi of commerce was made by Professor Baillon in 1875, who, from the material available, considered that to the plant mentioned at the head of this paragraph must be referred the bulk of commercial Jaborandi, a quantity also being afforded by *P. Selloanus.* Jaborandi has obtained a reputation as a very energetic diaphoretic sialogogue.

Piper methysticum.—The roots of this plant, which belongs to the natural order Piperaceæ, have been used from an early period in the Society and South Sea Islands, under the name of KAVA, in the preparation of a well-known intoxicating beverage. In 1876 the plant began to attract some attention as to its medicinal properties, since which time many experiments have been made to determine its physiological action. It has since been used in

practice in urethritis, leucorrhœa, dysuria, and all inflam
matory conditions of the urinary passages. In the Colonial
and Indian Exhibition, 1886, a spirit was retailed at the re-
freshment bars which was distilled from the roots of the
Kava plant. It was sold under the name of Kava
Schnapps or YANGONA.

Plantago ovata.—This plant is perhaps better known
under the name of *P. Ispaghula,* by which it was described
by Roxburgh. It is an annual belonging to the order
Plantaginaceæ, and found wild in North-Western India, and
also cultivated for the sake of its seeds, which are oval and
boat-shaped, of a greyish-pink colour, and extremely muci-
laginous ; by placing them in water they will yield a thick
mucilage, which is highly valued in India for its demulcent
properties, under the name of SPOGEL or ISPAGHUL seeds.
They were first introduced to notice in this country in 1870,
and have since been used in coughs, colds, diarrhœa, etc.,
as well as for feeding poultry.

Podophyllum peltatum.—A perennial belonging to the
natural order Berberidaceæ, and found growing in moist
shady situations all over the eastern side of the North
American continent from Hudson's Bay to New Orleans
and Florida. The properties of the root or rhizome have
long been known to the North American Indians, and it
has been used as a purgative in American Pharmacy since
1820, but it was not till 1864 that it was admitted to the
British Pharmacopœia. The active principle of the root,
under the name of PODOPHYLLIN, is now manufactured on a
very large scale both in America and England.

Rhaphidophora vitiensis.—This plant, which belongs to
the natural order Aroideæ, is supposed to furnish the
principal component part of the celebrated medicine TONGA.
This medicine was introduced to notice in 1879 as a new
drug from Fiji, having a high therapeutic value as a remedy
for neuralgia. The history of its introduction is singular,

and is as follows :—An Englishman who brought it here obtained it from a settler, married to a half-caste Tongan, who first prepared it for the use of her husband, and gave to the drug the name of her native island. The material as received here consisted for the most part of a mixture of broken pieces apparently of a root and some pieces of bark. From microscopical examination they appeared to belong to some aroideous plant, which was confirmed at a later period by the receipt of leaves, by which means the genus was established as *Rhaphidophora*, probably *R. vitiensis*. Tonga has been proved very efficacious in neuralgia, and may be included amongst the important medicines of the age.

Rhamnus Purshiana.—Under the name of CASCARA SAGRADA (lit. Sacred Bark), considerable attention has been paid since 1883 to the bark of the above-named tree or shrub, which belongs to the natural order Rhamnaceæ, and is a native of the Pacific slopes of North America, in which country it has become very generally used as a purgative. Introduced into this country from the United States in 1879, it has also found considerable favour with us.

Sarcocephalus esculentus. — A West African tree, belonging to the natural order Rubiaceæ, the bark of which, under the name of DOUNDAKÉ, has attracted much attention since 1885, both on the Continent and in this country, as a tonic and febrifuge, as well as for the golden-yellow colour contained in the bark. The fruit is known as the SIERRA LEONE PEACH.

Simaba Cedron.—The CEDRON, which is the name under which this plant is commonly known, is a small tree, 12 to 15 feet high, with a trunk measuring about 6 inches through. It belongs to the natural order Simarubeæ, and is a native of New Granada, on the banks of the Magdalena. It bears an oval fruit, about the size of a swan's egg, containing usually not more than one seed, about two inches long and half an inch broad. The earliest notice of the Cedron is contained in "The History of the Buccaneers," published in London in 1699, where its use as an antidote for snake bites is referred to. The method of using it is as follows:—When a person is bitten, a small portion of the seed, mixed with water, is applied to the wound, and about two grains scraped into brandy, or even into water, is given internally. This treatment is said to be an almost certain cure for the bites of the most venomous snakes, scorpions, centipedes, and other noxious animals. The seeds have an intensely bitter taste, and are said to have proved valuable in cases of intermittent fever. The seeds were introduced to Kew in 1846, and first brought to notice as a remedy in 1850, and again as a febrifuge and cure for toothache in 1884.

Strophanthus.—There are but few drugs of recent introduction that have attracted so much attention chemically and physiologically as the active principle of the seeds of the several African species of *Strophanthus*.

In 1870 Professor Fraser first pointed out, in a paper read before the Royal Society of Edinburgh, the powerful action of *Strophanthus hispidus* upon the heart, and stated, as the result of his experiments, that "it acted in a powerful and direct manner upon the cardiac muscular fibre, greatly prolonging the contraction of those fibres, rendering it continuous, and only to be overcome when relaxation occurs as a natural consequence of post-mortem decomposition." For some time after the publication of this paper little or nothing was heard of *Strophanthus*, till in 1877 it formed the subject of a paper by Messrs. E. Hardy and N. Gallois in the *Bulletin de Thérapeutique et Chirurgicale*. In 1885 it was again brought forward by Professor Fraser at the British Medical Association meeting held at Cardiff, and the paper was published in the *British Medical Journal* for November 14th, 1885. The publication of this paper naturally resulted in the attention of the whole medical profession being drawn to this new and important drug, and consequently there arose a very great demand for it—a demand, indeed, far exceeding the supply. Immature fruits containing unripened seeds, and consequently less powerful action, arrived in the market together with the seeds of other species than *S. hispidus*, so that the tincture prepared from them could not be relied upon. Of late, however, a better system of collecting seems to have been established, and tincture and tabloids of *Strophanthus* are now advertised as regular articles of trade. Though it was to *Strophanthus hispidus* that the credit was first given as possessing the valuable cardiac properties, a second species was described as *S. kombe*, which has since shared its reputation. These two are now included under the one species, *S. hispidus*. In Central Africa, the seeds when ground, mixed with water, and made into a paste, are used for poisoning arrows, both for purposes of the chase and in war.

Strychnos toxifera.—This plant is well known as furnishing the CURARE or WOURALI poison of British Guiana, which is prepared by scraping the bark, steeping it in water, and concentrating the fluid by evaporation. The natives use it for tipping their arrows in hunting as well as in war. It was brought to notice in this country in 1878 as a remedy in epilepsy, chorea, and hydrophobia, and is still included in our druggists' price lists.

Turnera diffusa, var. *aphrodisiaca.*—This plant belongs to a small order, Turneraceæ. A fluid extract of the plant was introduced to English pharmacy in 1874, under the name of DAMIANA, and recommended in renal and vesical diseases and in nephritic albumina. In some reports of its effects it is described as being "one of the best remedies in inflammatory diseases of the kidneys;" and taken as an infusion in the form of tea, prepared by pouring a cupful of hot water upon a teaspoonful of the dried leaves, it is said to have a marked effect upon sick headache.

CHAPTER VII.

OILS AND WAXES.

THE extended use of gas and the discovery of the petro-
leum or mineral oils during the last few years have had a
marked effect upon diminishing the use of vegetable oils as
illuminants. The spread of machinery, on the other hand,
has had an opposite effect in creating a demand for oil for
lubricating purposes; besides this there is always a large
demand for drying oils for mixing paints and for similar
uses. These facts, together with the increased use of oil-
cake for feeding cattle, cause a pretty brisk sale of oil seeds
generally, and oil crushers are alert and always ready to
give a trial to any new product of this nature arriving in
the English markets. A large quantity of these oil seeds,
especially those from the West Coast of Africa and Brazil,
find their way to the port of Liverpool, and it is surprising
how often new products of this nature, together with old
ones that have, perhaps, been sent years before and forgotten,
do come into that port. With a seed new to a broker,
coming into his hands for the first time, it is necessary that
he should make himself acquainted with its nature or properties—whether the oil it contains is wholesome or poisonous
—before he effects a purchase, it may be of a whole ship-load.
The nature of the seed governs not only the oil itself, but
also the marc or cake left after expression, which, in the
case of a sweet oil, would be valuable for cattle feeding,
while, on the other hand, in the case of a poisonous oil
it might bring about serious consequences.

The best-known oils, and those which are most largely
employed, especially in soap and candle-making—which take
the bulk of the oils imported—are COCOA-NUT and PALM

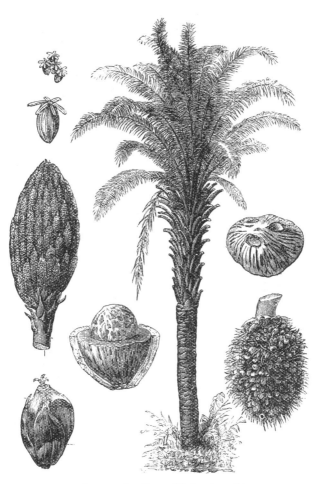

AFRICAN OIL PALM (*Elæis guineensis*).

OIL.   The first, it is well known, is the produce of *Cocos nucifera*, a widely-spread tropical palm ; and the second, the produce of *Elœis guineensis*, a palm confined to West Africa. The trade in both these oils has been largely developed since 1840, and is due, to a great extent, to the energy of Price's Patent Candle Company, which had its beginnings some fifty years or more since.   For some time the oil alone was imported, the cocoa-nut kernel being crushed in Ceylon, whence the bulk came.   Of late years, however, both oil and dried kernel have been imported, the latter known as " copra," which is submitted to pressure in this country.   So rapid did the utilisation of cocoa-nut oil become after the establishment of the company just referred to, that they turned out in the month of October, 1840, twenty tons of cocoa-nut candles, of the value of £1,590, and about twelve tons of stearic and composite candles, valued at £1,227.   In October, 1855, the quantity of stearic and composite candles made by the firm amounted to 707 tons, of the value of £79,500.   For the purpose of the general illumination on the occasion of Her Majesty's marriage in 1840, Price's Candle Company introduced a cheap candle that should require no snuffing, composed of a mixture of stearic acid and cocoa-nut stearine.   " The public, contrary to the general opinion of the candle-dealers, proved wise enough not to mind the candles being greasy, but as the light was good, the candles comparatively cheap, and the nuisance of having to snuff done away with, they received the new composite candles with great favour, and the manufacture rapidly grew."

In the development of the PALM OIL industry from *Elœis guineensis* a very important substance, namely GLYCERINE, was discovered ; it was first used in one of the hospitals for skin diseases in 1844.   Its uses at the present time are very numerous, and are well known.   About the year 1848 night-lights were introduced, and in the following year the well-known " Childs' Night-Lights " began to be made in large quantities.

The following are the returns of cocoa-nut and palm oil for the years stated between 1847 and 1889 inclusive :—

COCOA-NUT OIL.			PALM OIL.		
1847	- -	48,320 cwts.	1847	- -	366,840 cwts.
1857	- -	207,239 ,,	1857	- -	854,791 ,,
1867	- -	124,314 ,,	1867	- -	812,080 ,,
1877	- -	194,052 ,,	1877	- -	885,138 ,,
1886	- -	156,667 ,,	1886	- -	993,091 ,,
1887	- -	183,766 ,,	1887	- -	966,536 ,,
1888	- -	197,773 ,,	1888	- -	955,369 ,,
1889	- -	213,470 ,,	1889	- - 1,019,077 ,,

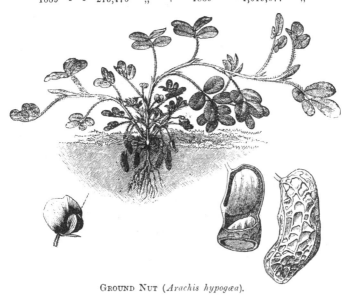

GROUND NUT (*Arachis hypogæa*).

GROUND NUT (*Arachis hypogæa*).—This is a diffuse herbaceous annual, growing one or two feet high, unknown in a wild state, but now much cultivated for the sake of its oily seeds in all tropical and sub-tropical countries, especially in West tropical Africa. After the fall of the flower the young pod pushes its way beneath the surface of the earth, where it ripens. The introduction of the ground nut as an oil seed into European trade dates from 1840, since which time the imports have increased enormously. There are no authentic records of the imports of ground-nut oil, but West Africa, India, and China supply by far the largest bulk. The oil is very free from stearine, and is consequently much used in pharmacy in the same way as olive oil, especially in India. With us it is also largely used for culinary and industrial purposes, as soap-making, etc.

COTTON SEED.—The cotton seed of commerce is furnished by several species of *Gossypium*. The seeds were first imported into the English market as oil seeds some thirty or forty years since, but it is quite within recent years that the trade in cotton seeds has assumed a position of importance. In America at the present time it has taken the place of a distinct industry, over 400,000 tons of seeds being annually expressed, the quantity indeed increasing every year. A large quantity of this oil comes to this country directly and indirectly. Egypt also sends cargoes of seeds to English ports for expression here. Much of the oil is used by soap-makers, besides which it makes a good lubricating oil; and when carefully refined in France, and put into white glass bottles, it is sent into this country as " Pure Olive Oil," and used for culinary purposes. So recently as December, 1888, the British Consul at Venice, reporting on the trade and commerce for that port for 1887, says that the action of the Italian Government in enacting a higher import duty on cotton oil with the intention of preventing its being mixed with olive oil has had a contrary effect, the price of olive oil being considerably lowered, the reason of which is said to be that by the mixture of cotton oil the ordinary qualities of olive oil, produced in the South of Italy, find an easier and more profitable sale. The residual cake, after the expression of the oil, is used for feeding cattle and as a fertiliser for the land.

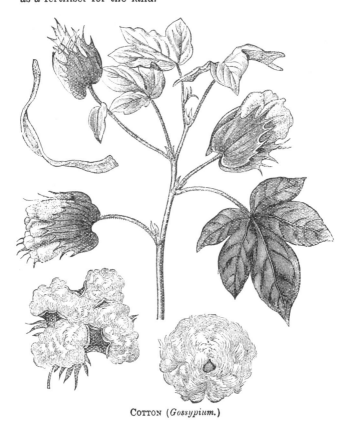

COTTON (*Gossypium.*)

DIKA or UDIKA FAT.—Under the name of DIKA BREAD the compressed seeds were first exhibited at the Paris Exhibition in 1855 as the produce of *Mangifera gabonensis*.

In 1859 it was brought to the notice of the Pharmaceutical Society, and in 1862 a report of its nutritive value was published in the *Journal* of the same Society ; from this it would seem that its composition is analogous to coffee, tea, cocoa, etc. ; and it was then suggested that it might become an article of commerce in this country. The substance is composed of the fatty kernels of the seeds of *Irvingia Barteri*, a simarubeous tree of West tropical Africa, and is made into masses of a cone-like form, sometimes weighing as much as fifty pounds. It forms an important article of food amongst the natives, and contains 70 to 80 per cent. of solid fatty matter. It has quite recently been stated that this fat is now used, mixed with cocoa butter from *Theobroma cacao*, for medicinal purposes.

Telfairia occidentalis.—A climbing plant belonging to the order Cucurbitaceæ, native of West tropical Africa, where the plant is cultivated for the sake of its seeds, which contain a sweet bland oil. They are cooked and eaten by the natives, and are said to be very palatable. The seeds are occasionally brought into Liverpool as oil seeds. The plant, which flowered at Kew in 1876, was raised from seeds received in 1870 from the Liverpool Botanic Garden.

Myristica angolensis.—A native of Angola, where it is known as MUTUGO. The seeds, which are about three-quarters of an inch long and half an inch broad, are ruminated like an ordinary nutmeg, but have no aroma and but little or no taste. They are said to contain about three-fourths of their weight of fatty oil. They were first imported into Liverpool as oil seeds in 1884.

Other species of *Myristica* to which attention has been directed as oil seeds are :—1. *M. surinamensis*, imported into Liverpool from Para as oil seeds in 1881. Like the former, they have no smell and very little taste. They are nearly globular, about the size of a small marble, and are known as CUAGO nuts by the Spaniards. They are said to yield 72 per cent. of their weight of solid fat, and the meal, after the expression of the oil, is described as useful as a substitute for linseed in making poultices.

2. *M. guatemalensis.*—A native of Guatemala, the seed of which is ovoid, about one inch long and half an inch broad. This also yields a solid fat in large quantity.

Hyptis spicigera.—An herbaceous plant belonging to the natural order Labiatæ. The small black seeds contain a large quantity of oil, and are occasionally imported into Liverpool from the west coast of Africa. They made their first appearance in 1883.

Polygala rarifolia.—A shrubby plant belonging to the natural order Polygaleæ, native of West Africa, about Sierra Leone and Angola. The seeds are very oily, and were first received at Liverpool in 1884 under the name of MALUKU seeds.

Lophira alata.—Under the name of MENI these seeds have recently been brought into Liverpool from West Africa for the sake of the oil they contain. The plant belongs to the natural order Dipterocarpeæ, and is known in Sierra Leone as LAINTLAINTAIN, where the oil is used, as well as in Senegambia, for cooking and for anointing the hair.

Pentaclethra macrophylla.—OWALA of the Gaboon, OPACHALA of the Eboe country. This is a leguminous tree growing to a height of 50 or 60 feet, the large seeds of which are used as food on the Niger, and the oil which is expressed from them in large quantities is used for domestic purposes, for lubricating machinery as well as for soap-making. The seeds are not a regular article of trade, but are occasionally imported into Liverpool.

Lallemantia iberica.—A plant belonging to the natural order Labiatæ, and said to be cultivated to a considerable extent from Syria to Northern Persia. The small seeds contain a very large quantity of sweet limpid oil, suitable

I

for culinary or other purposes. It was introduced to notice in England in 1880.

Under the names of M'POGO nuts, MABO nuts, and NIKO nuts, the hard bony fruits, *minus* the fleshy coverings in which they are enveloped when fresh, come occasionally into the port of Liverpool from the west coast of Africa, chiefly from Liberia and the Gaboon. The fruits of the M'pogo, which are imported from the Gaboon, are about two inches long and from one to one and a half inches in diameter. They contain three or four small roundish seeds, from which a very large percentage of oil can be expressed. The Mabo fruits are of an oblique-ovoid form, two inches or more long, and about an inch in diameter, with a very rough or channelled surface. The seeds of this kind are also very rich in oil, of a very fluid character. These fruits and seeds are imported from Liberia. The Niko nuts, which come also from Liberia, are of a similar bony nature, about two inches long and one and half inches in diameter. The seeds, like the other kinds, contain a large proportion of oil. Neither of these have become established articles of trade, though the oil seems to be of a character that might become useful. Owing to the want of authentic specimens of the leaves and flowers of the plants producing these fruits, they have never been botanically identified, though it has been surmised that they might belong to the genus *Parinarium*, of the natural order Rosaceæ. Probably, however, they may prove to be a species of *Elæocarpus*. They first made their appearance in Liverpool some ten or twelve years since.

CHAPTER VIII.

GUMS, RESINS, AND VARNISHES.

In the matter of new products coming under either of the above headings, there is but little to record as the discovery of the period under review. The points of interest connected with these substances lie more in the clearing up of doubts relating to their botanical origin and their accurate determination, as well as in their increased consumption and imports. The former, however interesting though it be, does not come within our scope, except where it bears on the development of the substance from a commercial point of view, or is instrumental in opening up new sources of supply. Under these circumstances our notes in this section will be necessarily limited. In passing, however, it may perhaps be of some interest to note that of Gum Arabic (which may be taken as the most important of the true gums), the imports have increased from 25,289 cwts. in 1839 to 75,399 cwts. in 1886, falling again in 1887 to 46,443 cwts., a decrease due to the disturbed state of the Soudan, whence the best kind of gum is obtained, and rising again in 1889 to 65,368 cwts. In consequence of the Soudan war, however, higher prices have ruled the market, and other gums have been brought into competition, the most notable of which is that which appeared early in 1888, under the name of Brazilian Gum Arabic. In appearance it resembles the ordinary quality of Gum Arabic, and is said to be derived from the Angico tree of Brazil, which has been described as *Acacia angico* of Mart. It is referred to in the *Kew Bulletin*, No. 17, for May, 1888, but since then the plant has been proved to be *Piptadenia macrocarpa*, Benth. It may also be worth while pointing out that the fragrant gum-resins, known as Balsams of Peru

I 2

and Tolu, were, fifty years ago, considered to be the produce
of the same tree, Tolu being the resin hardened by exposure.
It has, of course, been known now for some time that
Balsam of Peru is the produce of *Toluifera Pereiræ*, native
of San Salvador, Central America, while Balsam of Tolu is
furnished by *Toluifera Balsamum*, a native of Venezuela
and New Granada. Under the trade names of ANIME or
COPAL several kinds of hard fossil resin have long been
known in commerce, partly derived from Africa, and partly
from the East. The sources of these gums, which were then,
as now, used exclusively in the manufacture of varnishes,
were for a long time quite unknown. Indeed, the most
valuable resin—namely, that known as Anime—was until
comparatively recent times supposed to be the produce of
India, being shipped to this country from Bombay. It is,
however, now known to be furnished by *Trachylobium
verrucosum*, a leguminous tree of Zanzibar, the resin
being shipped thence to Bombay, and from Bombay to
England. The clearing up of this interesting point in
Economic Botany is due to Sir John Kirk, British Resident
at Zanzibar, who communicated a paper on the subject to
the Linnean Society in 1868, and sent full herbarium specimens of the plant to Kew, as well as a fine series of gum
specimens. Seeds of the tree have since then been introduced into India and Australia.

The best Anime is that which is dug from the ground
near the roots of the trees, or where the trees once stood,
but have now disappeared. Regarding the export of Anime
from Zanzibar, Sir John Kirk says it sometimes reaches
800,000 pounds, of the value of £60,000.

What promised to be a very important source of Copal
was made known in 1883, when the British Consul at
Mozambique reported the discovery at Inhambane of a tract
of Copal forest fully 200 miles long. Samples of this new
fossil Copal or Anime were sent to England, and upon

practical tests being made upon its suitability for varnish-making, was favourably reported on, and valued at from £80 to £100 per ton. Some of these samples are now contained in the Museum at Kew. It is the produce of *Copaifera Gorskiana.*

Some later information on the subject is given in the *Kew Bulletin*, No. 24, for December, 1888, where an extract from a letter from Inhambane, under date Feb. 5, 1886, states :—" Many tons of copal have been exported from Inhambane. For some choice pieces I have received as high as £13 10s. per cwt. The average price realised on larger lots has been £7 per cwt. The forest containing the trees extends from the River Sabia in a south-westerly direction as far as Beleni."

Fresh seeds were also received at Kew, and several hundred plants raised from them, which have been distributed from Kew to India, Fiji, Singapore, Jamaica, Trinidad, Demerara, Dominica, and tropical parts of Australia.

Another varnish-making resin, but little less important than Zanzibar Anime, is Kauri or Cowdie resin of New Zealand. This, like Anime, is a semi-fossil resin, more commonly known in trade, however, under the name of Kauri *Gum*, and is the produce of *Dammara australis*, a very large coniferous tree valued alike for its timber as for its gum. (*See* Timbers.) The best Kauri gum is dug from the ground beneath the trees, or where the trees do not at present exist. Thirty-three years ago Kauri gum was imported into this country only in small quantities, for we find that in 1853 the total exports of the gum from New Zealand to all countries amounted to only 829 tons, of the value of £15,971 ; in 1883 this had risen to 6,518 tons, valued at £336,606. It is said that over two-thirds of the produce goes to the United States ; and there are no available returns of the imports into this country, though the quantities are very large. Though gum-digging gives employment to a large number of persons, they generally consist of the lowest orders. Quite recently, however, it has been stated that, in consequence of depression of trade in New Zealand, a large number of men have taken to Kauri digging, as many as 10,000 being so occupied at present, and the quantity of gum brought to the Auckland market has very considerably increased.

Under the name of OGEA GUM a hard fossil resin of the copal character was introduced to notice in 1883 by Captain (now Sir Alfred) Moloney from the Gold Coast. It is described as being the produce of a leguminous tree closely allied to *Daniellia thurifera;* for lack of proper material, however, its species has not yet been determined. The gum is used by the natives both for lighting fires and for illuminating purposes ; powdered, it is also used as a body perfume by the women. It exudes from the trunk either from wounds or from holes caused by the boring of insects. The gum has not yet appeared in commerce.

CHAPTER IX.

THE greatest development in the direction of dyes has not been towards those of vegetable origin. On the contrary, for the last twenty or thirty years vegetable dyes have been gradually displaced by the advances of chemical science in utilising coal-tar, and in the artificial preparation of colouring matters to supersede the old vegetable dyes. In this direction we need but refer to the serious blow given to the trade in Persian berries (*Rhamnus infectorius*) in the Levant by the discovery of the Aniline dyes, or to the more recently threatened substitution of chemically - prepared Indigo for that of vegetable origin. So alarming did this discovery seem to be to the indigo-planters in India that we cannot refrain from quoting the following paragraph from a letter of Professor Armstrong published in the Kew Report for 1880. He says :—" Notwithstanding the number of operations involved in the manufacture, it is stated that it will be possible thus to produce indigo at such a price that it can even enter into competition with the natural article, and that by substituting the method of dyeing previously described for the troublesome and somewhat uncertain indigo vat method, there will be a still more distinct advantage gained over the natural article. It is difficult at present to estimate the influence which this discovery may have on the production of Indigo in India; but when it is remembered, to take an analogous case, that the discovery of a process of manufacturing madder red was only made in 1869, and that now it is almost impossible to procure natural madder red or garancine, the annual value of the imports of which into the United Kingdom alone for the years 1859 to 1868 amounted to about £1,000,000 sterling, it is difficult to avoid the conclusion that artificial indigo will most seriously interfere with, even if it does not within a very few years altogether displace, the natural article."

Though this was written ten years ago, vegetable indigo still retains its position in the market.

The hard, dried fruits now imported from India in such large quantities under the name of MYROBALANS were only just making their way into commerce when Her Majesty ascended the throne; at the present time they come into this country from India for the use of tanners to the extent of over 640,000 cwt. a year. Two kinds are known in commerce—the CHEBULIC MYROBALAN (*Terminalia chebula*) and the BELLERIC MYROBALAN (*T. belerica*).

In 1875 the pods of a leguminous tree of South America (*Cæsalpinia brevifolia*) were introduced from Santiago under the name of ALGAROBA. They were said at the time to contain a large amount of tannin—90 per cent.—and to be superior even to DIVI-DIVI (*Cæsalpinia coriaria*). In 1878 some pods of *Wagatea spicata* were sent from India to test their value for tanning purposes. They were said to contain 15 per cent. of tannic acid. The plant is a native of the Concan, and is a scrambling thorny shrub belonging to the natural order Leguminosæ. Seeds of this plant were distributed from Kew to Demerara, Dominica, Jamaica, Trinidad, and other places.

Elephantorrhiza Burchellii.—Under the name of ELANDS BONTJES the root-bark of this leguminous plant first attracted attention in 1866, when a paper was read before the Pharmaceutical Society by Professor Attfield, and published in the *Pharmaceutical Journal*, Vol. VIII., 2nd Series, p. 316. The plant, which was there referred to a species of *Acacia*, is said to furnish food from its seeds, a medicinal infusion from its root, and also a valuable tanning material. It was found upon analysis to contain 20 per cent. of tannic acid. Nothing further was heard of this root till 1886, when it was exhibited in the Natal Court of the Colonial and Indian Exhibition. Mr. T. Christy, in his *New Commercial Plants and Drugs*, No. 10, published in 1887, says :—" Mr. W. N. Evans, who tested the root, states that it contains 25·37 per cent. of tannin, and that, if it were to work up in a similar manner to Mimosa bark, the best samples might be worth from £14 to £15 per ton. With regard to its practical value as a tanning material for leather, from the incomplete trials that were made with the small quantity received, it was found to give too red a colour; but I should not like to speak positively upon this point, as in treating a few cwts. of the roots at a time it might be found that this detriment could be overcome."

Phyllocladus trichomanoides.—A very large coniferous tree of New Zealand, where it is known as TANEKAHA. The bark, which is of an orange-yellow colour, has of late years come largely into use in this country for dyeing kid or dogskin gloves the fashionable golden orange.

Under the name of CANAIGRE a tanning material has been known in America for the last ten years or more, and accounts of it have appeared from time to time in this country. In the *Leather Trades Circular* for August 8, 1885, under the head of "New Tanning Agents," the following occurred :—" An Arizona paper states that a new tanning agent, likely to be of great value, has been discovered, one which also has the property of adding weight to the leather. The plant is an annual, and grows upon desert and dry upland soil. It is known by the Mexicans and Indians as GOUAGRA. . . . Practical use demonstrated that the tanning properties of this root were about three times as great as the common oak bark, and that in all essentials it was superior to the bark in the manufacture of leather." The roots, which are fleshy, are from three to six inches long, and one and a half to three inches broad, of a somewhat oval shape, and covered with a dark brown skin.

The stems and leaves are described as being acid, like
rhubarb, and are used in a similar way in California and
Utah under the name of WILD PIE plant. In Texas the
roots are used for tanning. The plant has recently been
determined as *Rumex hymenosepalum*, belonging to the
natural order Polygonaceæ, and from an analysis made so
recently as March of the present year (1890), it would
seem "that these roots will be a valuable addition to our
list of tanning products." This interesting substance is
fully detailed in the *Kew Bulletin* for April, 1890.

CHAPTER X.

PAPER MATERIALS.

THE enormous demand for paper that has sprung up of late years has, like the demand for so many other products, caused those most interested to divert their attention to new sources of material. Notwithstanding that so long ago as 1801 Matthias Koops obtained a patent for manufacturing paper from hay, straw, thistles, waste, and refuse of hemp and flax, and different kinds of wood and bark, linen and cotton rags remained almost the exclusive material from which paper was made until about forty years ago. Then, as the penny daily paper appeared and became general, old ropes, sacking, jute, and a host of other substances were pressed into the service. The most important introduction, however, was Esparto (*Stipa tenacissima*). It was in 1856 that the late Mr. Thomas Routledge, so well known in connection with the paper trade, obtained a patent for manufacturing paper from Esparto grass. Some of the first paper made from this grass was used for printing the Report of Dr. Forbes Royles' paper on Indian Fibres, and formed the number of the *Journal of the Society of Arts* for November 28, 1856. In that year the total imports of Esparto amounted to only 50 tons, the whole of which was worked up at Mr. Routledge's mill at Eynsham, near Oxford. In 1864 the quantity imported rose to 50,000 tons, and in 1886 the return was over 200,000 tons, which continues to be the average quantity imported at the present time. This enormous demand for Esparto, coupled with the destructive manner of collecting it—namely, by tearing it from the roots—has considerably diminished the sources of supply, so that at present paper-makers are as much alive as ever to new materials. Another substance to the utilisation of

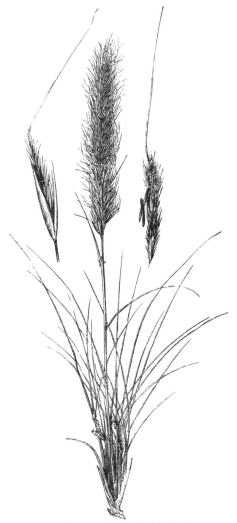

ESPARTO (*Stipa tenacissima*).

which Mr. Rout-
ledge paid con-
siderable attention
was the young
stems of bamboos,
and he succeeded
in showing that
a very fine paper
could be made
from these stems,
as he published a
pamphlet on the
subject in 1875
which was printed
on paper made
from bamboo. It
was proposed that
the stems should
be crushed at the
place of growth
and shipped in
that state to this
country, so as to
reduce bulk and
freight. The in-
terest in the pro-
ject to utilise the
bamboos as a
paper material be-
came general, both
in England, India,
and America. As
these gigantic
grasses will grow
in almost any tro-

pical country, there seemed to be an unlimited source of supply. Opinions of botanists, however, were divided as to whether the continued cutting of the young bamboo stems would not considerably weaken them and eventually reduce the supply. One thing to be borne in mind in considering this subject is that the several species of *Bambusa* and *Dendrocalamus* are equally suited to the manufacture of paper, and that in India bamboos are very plentiful. Indeed, Sir D. Brandis states "that there are about 1,800 square miles of almost pure bamboo forest in the Arrakan division of British Burma within a moderate distance from the coast, and all accessible by navigable streams." The accessibility is, of course, a matter of considerable importance in bringing bulky material down to a point whence it could be the more readily transferred for shipment to England.

Notwithstanding the interest taken in the Bamboo as a probable source of paper material, it has not, down to the present time, become a recognised article of trade.

BAOBAB (*Adansonia digitata*).—The fibrous bark of this West African tree was first brought to the notice of the paper-maker in 1876. It was proved upon trial to possess all the necessary properties for making an excellent paper. The drawback to its general utilisation has been the slow growth both of tree and bark, and the probability of a failure in the supply.

PAPER MULBERRY (*Broussonetia papyrifera*).—This well-known tree, from the bark of which the Polynesian islanders make their Tapa cloths, and the Japanese a large portion of their excellent paper, which they put to such a multitudinous variety of uses, was first brought to the notice of English paper-makers in 1879. The late Mr. Routledge, in reporting upon it, described it as "nearly, if not quite, the best fibre he had seen," requiring very little chemicals, and giving an excellent yield.

In Japan the plant is cultivated for paper-making, the young shoots being used for the purpose.

CALIFORNIAN " CACTUS."—Under this name the stems

PAPER MULBERRY (*Broussonetia papyrifera*).

of a plant were brought to the notice of the Kew authorities in 1877 as a valuable paper material. It was difficult, from the material first brought, to determine its botanical affinity. In 1878, however, further material came to hand, from which the plant turned out to be *Yucca brevifolia*, described in California previously, but incorrectly, as *Yucca Draconis*. Forests of this plant existed in the Mohave desert for several miles, through which the Southern Pacific Railway runs. The stem of this plant, which grows to a diameter of a foot or more, is of a very fibrous character, and it was soon found to be an excellent paper material, in consequence of which the plants have been systematically cut down and turned into paper, which was at one time used almost, if not quite, exclusively for printing the *Daily Telegraph* upon.

Amongst other vegetable substances more or less suitable for paper-making that have been brought to notice at Kew during the past ten years may be mentioned the following :—

Cavanillesia platanifolia.—A plant belonging to the Malvaceæ, found abundantly in the eastern part of the State of Panama, and as far east as Carthagena, known as VOLANDERO. The fibrous bark was found to pulp well, bleach readily, and to make a strong opaque white paper of fine quality. This was tested in 1877.

Uniola virgata.—A grass locally abundant in Jamaica. In consequence of its bulky nature it would not pay to send it in its raw state to England, but it might be exported in the form of paper stock, and form a somewhat inferior substitute for Esparto. It was tested in 1876.

Calotropis gigantea.—Under the name of MUDAR this asclepiadaceous plant is well known in India, where the fibre from its stems is used in making cordage, and the floss from the seeds for stuffing cushions, and occasionally for weaving. It was first proposed as a paper material in 1877, but the trials made with it were not satisfactory. Again, in 1880 it was spoken favourably of in a report from India, but it has still not been received favourably in this country.

Ischæmum angustifolium.—This is the BABAR or BABOI Grass of India, and grows abundantly in many parts of the country. It has long been used in India for making into ropes and cordage, and has latterly become one of the principal paper materials, being largely used in the Bally Paper Mills near Calcutta. It was introduced to notice in England in 1878, and Mr. Routledge reported upon it as follows :—" A small quantity of bleach brings it up to a good colour. The ultimate fibre is very fine and delicate, rather more so than Esparto, and of about the same strength ; the yield, however, is 42 per cent., somewhat less. I think I may venture to say it will make a quality of paper equal to Esparto."

The great drawback to the general utilisation of the fibre in this country is that the plant has to be collected in India over wide and distant areas, and its bulky nature increases the cost of freight. It might, however, be converted into paper stock in India, and exported in that form. That the plant is capable of extended cultivation in India if a de- mand for it should spring up in this country, has been recently shown in an account of its culture published in the *Proceedings of the Agricultural and Horticultural Society of India* for October, 1887. The plant is, perhaps, equally well known under the names of *Eriophorum comosum* and *Pollinia eriopoda*, under both of which it has been de- scribed.

Molinia cœrulea.—This well-known British Grass was brought to notice as a probable source of paper material in 1878, and in the Kew Report for 1879 it is stated :—" Mr. N. G. Richardson, of Tyaquin, county Galway, has actively promoted its experimental cultivation in the West of Ireland. At a private meeting held at Athenry, a com- mittee was formed to raise subscriptions to plant ten Irish acres of bog with it at Tyaquin. Mr. W. Smith, of Golden Bridge Mills, had manufactured paper from this grass with which he was so well satisfied that 'he was prepared to buy 1,000 tons if any one would supply him.'"

Secale cereale, RYE STRAW, was proposed in 1879, Mr. Routledge's report being that " it is very largely used in the States, also on the Continent. It will make a harder and firmer paper than any other cereal straw, except perhaps maize."

Musa, spp.—The utilisation of Plantain and Banana stems for paper-making was brought forward in the Kew Report for 1881. It is there pointed out that there can be no question as to the suitability of the fibre for the purpose, but that the practical difficulty has been in dealing with the 90 per cent. of water which the stems contain. By mechanical treatment, however, the fibre of a plantain stem can be dried off within a period of eight hours, and as the plants are very abundant in India and Burma, it might be worth while to systematically extract the fibre for paper-making. Dr. King, of Calcutta, reporting on this subject, says :—" In my opinion this proposed plantain industry has a good deal of promise about it, and I think it might be well worth while for Government to spend a little money in sending a sufficiently large shipment to the London market, and to allow it to be sold for what it will fetch in small lots, so that the new material may become generally known to the paper-making interest. If the fibre answers for paper, Government need do no more ; the matter will, no doubt, be taken up by private enterprise.

" The Bengal Government will be prepared to give all reasonable assistance to any mercantile firm or individual wishing to try experiments, and will arrange for future supplies at reasonable rates. It will also give such other assistance as may be deemed necessary and proper."

Commenting on this, Sir Joseph Hooker says:—"Whatever the success of the enterprise in India, I think the matter is well worth attention in the West Indies. The

J

cultivation of bananas for export is assuming a constantly

PLANTAIN (*Musa paradisiaca*).

increasing magnitude. Each banana stem is useless after it

has borne fruit, as it does not do this more than once. To work up the decaying stems into paper pulp, if it could be done inexpensively, would be a desirable addition to the profit of banana-growing, and would get rid of the evils incident to the decomposition of the useless stems."

WOOD PULP.—The reduction of the trunks of certain coniferous trees, as well as of the Poplar, in the preparation of wood pulp, is a well-known industry of Norway and Sweden, where factories for this purpose are still increasing, and whence a large portion of the product finds its way to this country. It is a comparatively new industry, and one capable of almost unlimited extension.

J 2

CHAPTER XI

FIBRES.

FEW branches of manufacture have attracted so much attention in recent years as the application of new fibres. The numerous uses to which fibres are put will sufficiently explain this; paramount, of course, must always be that for textile purposes, then for rope and cordage, next as a substitute for bristles in broom and brush-making, and finally for paper-making, which has been treated of under a distinct heading.

It is, then, for the first three uses that we have now to consider the fibre supply; and in glancing at the subject from its first aspect, mainly as furnishing textiles, we may briefly allude to the cotton supply, which in 1800 was only about 600,000 cwt., the increase going on steadily down to our own time, as will be seen from the following statistics :—

1837	Total imports of raw Cotton	...	3,636,489	cwt.
1856	,,	,,	,,	...	9,141,842	,,
1860	,,	,,	,,	...	12,419,096	,.
1862	,,	,,	,,	...	4,678,333	,,.
1866	,,	,,	,,	...	12,295,803	.,
1886	,,	,,	,,	...	15,187,299	.,
1887	.,	,,	,,	...	15,903,117	.,
1888	,,	,,	,,	...	15,246,408	,,
1889	,,	,,	,,	...	17,159,316	,,

It will be remembered how seriously the American civil war affected the cotton trade in this country, and this is specially marked in the above table. Much larger supplies were at that time drawn from British India, and of the total imports for last year British India supplied 2,438,968 cwt.

In 1876 a new kind of cotton was introduced to the notice of planters under the name of BAMIA COTTON. It made its first appearance in Egypt, and attracted a good deal of attention on account of its mode of growth and its abundant fruit-bearing. It was described as sending off branches regularly from the bottom of the main stems upwards, but bearing close to the ground two, three, or more branches, and then rising to a height of eight or ten feet without a branch. This erect growth was considered an advantage, inasmuch as a much larger number of plants could be grown within a given area than is possible with ordinary cotton. The plant was also described as a prolific fruit-bearer, so that the yield was estimated at a considerably higher rate than any other known variety. In consequence of these very strong recommendations the seeds were distributed as widely as possible from Kew with very varied results. The quality of the cotton was reported as not to be materially different from that of ordinary Egyptian cotton, of which, indeed, it was found to be a fastigiate variety. Bamia cotton is now seldom or never heard of.

A textile fibre of undoubted quality which still awaits development is the so-called CHINA GRASS. This fibre seems to have made its first appearance in this country, in the form of finely-woven handkerchiefs, not long before 1849, for it was about this time that a specimen of the fabric was received at Kew together with other materials, from which it was found that the plant furnishing it, though called China Grass, was in reality a bushy-growing nettle—the *Bœhmeria nivea* or *Urtica nivea* of botanists. From this time the fibre began to attract much attention, and a patent was obtained in the same year (1849) in connection with its preparation. At the Great Exhibition in 1851 three prize medals were awarded for China Grass fibre. It was then proved that from the fibre, properly cleaned and prepared, fabrics could be woven equal in every respect to the finest French cambric. Notwithstanding this, the interest in China Grass dwindled down and remained in abeyance for some time, till in 1865 a fresh interest was given to it by the American Vice-Consul at Bradford, Yorkshire, suggesting to his Government at Washington the desirability of their introducing the plant and fostering its growth in the United States, for the double purpose of utilising its fibre in America and of exporting it to this country. The practical results of this communication, though it excited fresh interest in this country at the time, were almost *nil*. The great desideratum was the invention of a machine that would clean the fibre, and prepare it at such a cost that it might be put into the market at a price to compete with other textiles of a similar character; and with the hope of attaining this end, the Indian Government offered in 1869 prizes of £5,000 and £2,000 for such a machine. A Mr. Greig was the only competitor, and his machine did not altogether fulfil the conditions necessary for complete success, so that the matter again dropped. In the meantime the China Grass plant has been grown for experimental purposes in the South of France, near Marseilles, and in Algeria, and many new inventions in machinery for its preparation have been made in England, America, and on the Continent. During the year 1887 a fresh impulse was given to the fibre by a series of experiments with new machinery in Paris, as well as by the adaptation of a flax-cleaning machine, invented by Mr. Wallace, and exhibited during the year at an Exhibition of Irish Industries held in London. At a still later period— namely, in the *Kew Bulletin* for December, 1888—it is stated " that those who have in a measure been successful in preparing the fibre in commercial quantities are disappointed at the reception it has received at the hands of the spinners and manufacturers."

The extended cultivation of the plant presents no difficulties, given a suitable soil and a locality possessing the necessary climatic conditions of heat and moisture. There is no doubt that the Ramie or China Grass plant could

be cultivated in most of our tropical possessions. Regarding the question of the decortication of the stems, this
problem remains still unsolved. And on this, as the *Kew
Bulletin* says, "really hangs the whole subject. The third
stage [that of spinning] is disappointing and unsatisfactory
because the second stage [that of decortication] is still uncertain; and being thus uncertain, the fibre is necessarily
produced in small and irregular quantities, and only comes
into the market by fits and starts. It would appear that
Ramie fibre differs so essentially from cotton and flax that
it can only be manipulated and worked into fabrics by
means of machinery specially constructed to deal with it.
Owing to the comparatively limited supply of Ramie fibre
hitherto in the market, no large firm of manufacturers have
thought it worth while to alter the present or put up new
machinery to work up Ramie fibre. If appliances or processes for decorticating Ramie in the colonies were already
devised, and the fibre came into the market regularly and
in large quantities—say, hundreds of tons at a time—there
is no doubt manufacturers would be fully prepared to deal
with it. At present the industry is practically blocked
by the absence of any really successful means of separating
the fibre from the stems, and preparing it cheaply and
effectively. This, after all, is the identical problem which
has baffled solution for the last fifty years."

Further trials in cleaning Ramie fibre by machinery
were made in Paris during the Exhibition of 1889, the
results of which have been recorded in the November and
December numbers of the *Kew Bulletin* for that year. It
will suffice for our purpose to know that the conclu-
sions arrived at were that France appeared to be the best
market for the fibre. A well-known London firm of fibre
brokers, reporting on the trade in November, 1889, say
that strips of the bark—known as ribbons—were sold during
that week at from £14 to £16 per ton, and that they were

disposed to think that the bases of a real trade in the article were in process of formation.

The plant is a native of China, but is cultivated in India and the Malay Islands. By the Chinese it is known as Tchou-ma, in Assam as Rheea, and in the Malay Islands as Ramie. It has been introduced at different times into most of the British colonies.

About the year 1860 a substance called PINE WOOL was introduced to notice, two factories having been established near Breslau, in Silesia. The process consisted of reducing the pine-leaves to a coarse kind of fibre of a brownish-yellow colour. This was used for stuffing cushions, mattresses, etc., and as a kind of wadding; more recently it has been made into a yarn, and woven with animal wool, and sold as pine-wool flannel, which is said to have advantages over ordinary flannel, inasmuch as it keeps the body warm without heating, and is very durable. The pine chiefly employed is *Pinus Laricio*. More recently—namely, within the last two or three years—pine wool has been made in North America from the long leaves of the Turpentine Pine (*Pinus australis*), and used for making mats and carpets.

Perhaps no other fibre, whether textile or otherwise, has made such rapid strides as a commercial commodity as JUTE. The beginning of the Jute trade is intimately associated with Dundee, and dates back near upon fifty years. It is the inner bark of two or more species of *Corchorus*, of which *Corchorus capsularis* and *C. olitorius* are the chief. They are annual plants, belonging to the natural order Tiliaceæ, and are now largely cultivated in India, especially in Bengal, exclusively for the sake of this fibrous bark. This bark was at one time used only to make Gunny bags, in which to export Indian raw sugar; these, after being emptied of their contents in this country, were sold to the Jews, who extracted the remaining sugar by boiling, and then sold the old bags to the paper-makers, to be converted into pulp or paper stock. The fine glossy character of the jute fibre soon, however, began to recommend itself for textile purposes, and 9,300 tons were imported into this country in 1846, which rose in 1887 to 373,480 tons.

JUTE (*Corchorus capsularis*).

At first jute was only used for mixing with wools in cheap druggets and carpets. At the present time it is applied to a great variety of purposes, such as imitation tapestry, carpets, cords, twines, and even for mixing with cheap silks, to which it lends itself on account of its bright glossy appearance.

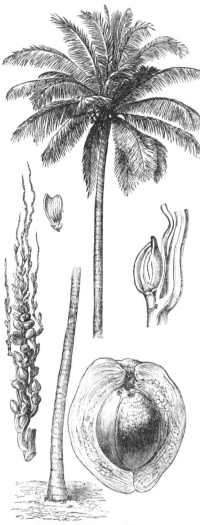

Amongst vege-
table fibres used
for brush and
broom - making,
several very im-
portant introduc-
tions have been
made, foremost of
which, of course,
is the fibrous husk
of the Cocoa-nut
(*Cocos nucifera*).
This fibre, now so
generally known
by the name of
Coir, has become
within the last
twenty or thirty
years a most im-
portant article of
import.   Its in-
troduction may be
said to date from
about the year
1836, when a shop
for the sale of
articles made of
Coir was opened
in Agar Street,
Strand.   In 1839
a partner in this
business took out
a patent for the
manufacture of
various fabrics

COCOANUT (*Cocos nucifera*).

from the fibre, and from this time its uses rapidly increased. One of the earliest applications of cocoa-nut fibre was for stuffing mattresses and cushions, and for this purpose it was cleaned, crushed, and dyed in imitation of horsehair ; for brushes it was also early applied, and is still largely used for this purpose. The treatment of the fibre and the improvement of the machinery used had so far developed in 1842 that on the occasion of the baptism of the Prince of Wales in that year, St. George's Hall, Windsor, was laid with cocoa-nut matting, which fact was recorded in the *Times* of January 26th in that year as follows :—"The floor was covered first with a matting made of the husk of the cocoa-nut."

The Exhibition of 1851 was the means of giving a further impetus to the trade, from which time it has gone on increasing. In the process of separating the fibre from the cocoa-nut husk three distinct commercial articles are produced—namely, the long fibres, used for matting and mats ; the shorter or more stubborn fibres, for brooms and brushes ; and the still shorter or refuse, for horticultural purposes. As a further illustration of the commercial importance of Coir, it may be stated that 5,246 tons were imported in 1866, which had increased in 1882 to 11,590 tons.

Another important brush-making material, but of more recent introduction, is BASS or PIASSABA, the produce of two distinct palms—namely, *Leopoldinia Piassaba* from Para, and *Attalea funifera* from Bahia. These two kinds are distinguished in trade, the fibre of the *Attalea* being superior to that of *Leopoldinia* for brush-making, on account of its being stiff and yet "springy," so that longer lengths can be used ; the Para fibre is more flexible, and can only be used in short lengths ; it is, however, of a brighter colour. The *Attalea* fibre can be obtained either very fine or very thick and strong ; each fibre is more or less round, while the Para kind is flat.

The introduction of Piassaba fibre into England for brush-making dates back about forty years, and is almost, if not entirely, due to the exertions of Mr. Arthur Robottom. When first introduced, it was used exclusively for road-

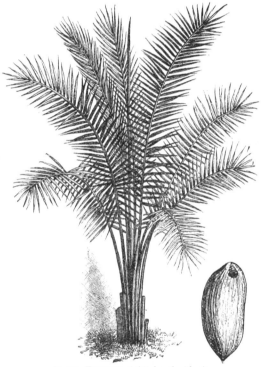

BAHIA PIASSABA (*Attalea funifera*).

brushes or brooms, which were generally known as whale-bone brooms; indeed, before the nature of the material was fully ascertained it was considered to be split whalebone. The fibre is easily collected from the trunks by cutting it away at the bases of the leaves; and when a sufficient quantity is collected, it is roughly cleaned by fixing pieces of stick in the ground with sharp points and drawing the fibre through them. This rough system of combing separates

KITTOOL, OR WINE PALM (*Caryota urens*).

the flossy fibre, which is not exported, but is used in Brazil for making ropes to tie poles together, or for making fences. The Piassaba is then pressed in bales or bundles ready for shipment. In 1861 nearly 6,000 tons of Piassaba were imported into England. The price has considerably advanced of late, owing to the diminished supply; and much interest is now being shown in the discovery of similar fibres from other sources. About eight years ago a new kind of Piassaba was introduced to the British market from Madagascar, and still forms an article of import. The fibres are thinner and much softer than those of either the Para or Bahia kinds, and consequently not so valuable for brush-making. Though this Madagascar Piassaba is proved to be the produce of a palm, the exact plant has not yet been determined. More recently a thick, whalebone-like fibre has been introduced as Lagos Piassaba, but little is at present known of it. It is apparently the produce of *Raphia vinifera*.

Another fibre that has recently taken a prominent position in the brush trade is KITTOOL, which is found in large quantities around the bases of the leaves of *Caryota urens*, a well-known Cingalese palm. Kittool fibre has been known in this country for some thirty or forty years, but it is within the last five or six years that it has become a regular commercial article. When first imported, the finer fibres were used for mixing with horsehair for stuffing cushions. As the fibre is imported it is of a dusky-brown colour; but after it arrives here, it is cleaned, combed, and arranged in long straight fibres, after which it is steeped in linseed oil to make it more pliable; this also has the effect of darkening it, and it becomes indeed almost black. It is softer and more pliable than Piassaba, and can consequently be used either alone or mixed with bristles in making soft, long-handled brooms, which are extremely durable, and can be sold at about a third the price of ordinary hair brooms. The use of Kittool fibre is said to be spreading not only in this country but also on the Continent.

Under the name of MEXICAN FIBRE or ISTLE a stiff fibre is now imported into the English market, chiefly for making scrubbing and nail brushes. The history of this fibre is

interesting, and may be given briefly as follows :—When
the war broke out between England and Russia, one of the
sources of hemp—namely, from Russia—was stopped ; the
Istle, which was known to some Mexican merchants, was
suggested as a substitute, and a small trial shipment was
made to England. It wàs soon found, however, that it was
unsuited for rope-making. A portion of it having come into
the hands of Mr. Robottom, whose name has before been
mentioned in connection with Piassaba, he at once suggested
its use for brush-making, and purchased the whole consignment of about twenty tons that had been shipped from New
York to Hamburg. On arrival in this country it was sold
for about £28 per ton ; the price soon rose to £85 per ton,
falling to £18, and afterwards rising again at the time of
the insurrection in Mexico to £140 per ton. The trade
afterwards increased very rapidly, and the fibre is now
imported in very large quantities, chiefly from Tampico, and
used for making scrubbing and nail brushes, whitewash
brushes, bath brushes, etc., and at one time it was largely
used by crinoline-makers. The source of this fibre was
unknown till in 1879 Dr. Parry sent specimens to the Kew
museum under the name of *Agave Lechuguilla.* This, however, has quite recently (*Bulletin of Miscellaneous Information, Royal Gardens, Kew, No.* 12, *December,* 1887,
p. 5) been shown to be identical with *Agave heteracantha,* to
which plant Mexican fibre or Istle must now be referred.
The present value of this fibre is stated to be about £26
per ton.

Another Mexican brush fibre, the botanical source of
which has been quite recently cleared up, is that known as
BROOM ROOT or MEXICAN WHISK. Though it appears to be
a comparatively new industry, there seems to be no record
when it was first introduced. It is shipped from Vera Cruz,
chiefly to Germany and France, a small quantity only
coming direct to this country. In France, however, it is

mixed with Venetian Whisk, the roots of *Chrysopogon Gryllus*, which, though somewhat lighter in colour, are similar in appearance, but of a superior quality ; and in this mixed condition it is exported to England for making clothes, velvet, carpet, and dandy brushes. The roots are known in Mexico as " Raiz de Zacaton," and are referred in the *Bulletin of Miscellaneous Information, Royal Gardens, Kew, No.* 12, *December*, 1887, p. 9, to *Epicampes macroura.*

About twelve years ago a new material was introduced for gardening purposes—namely, for tying plants—under the name of ROFFIA or RAFFIA ; for some time the origin of this article remained unknown, but it was subsequently proved to be the thin but very strong cuticle of the leaf of *Raphia Ruffia*, a palm, native of Madagascar. It is exported chiefly to Mauritius, and thence to England—at the present time in very large quantities. Its value in the London market ranges from £25 to £200 per ton, but its average price may be taken at from £40 to £50 per ton. In Madagascar this same substance, split into fine threads and dyed, is used for making mats and cloths, some of which are very beautiful.

It is said that the cuticle of the leaves of *R. tædigera*, a Brazilian species, is also exported to this country, and helps to make up the bulk of the Roffia of trade. This material has entirely supplanted the old Cuba Bast from *Hibiscus elatus*, which was so largely used in gardens about forty years ago. It was originally used for tying up bundles of real Havanna cigars ; and during the Russian war, when the bast from the lime-tree became scarce, it was used as a substitute for that article, and has now passed from notice to give place to Roffia.

In the London International Exhibition of 1862, amongst the South African products some prominence was given to a stiff black fibre which was advocated for brush-making, as a substitute for horsehair, and for paper-making. This fibrous substance was known as PALMITE or PALMIET, and is found in quantities encircling the stems of *Prionium Palmita*, a stout-growing plant of South Africa, belonging to the natural order Juncaceæ. Though it attracted some attention at the time, it never came into actual use.

Another substance which has come into use in recent years as a substitute for horsehair is known as CRIN VEGE-TAL, and consists of the crushed fibres from the leaves of *Chamærops humilis*, the European Fan Palm. It is cultivated in some parts of Southern Europe and Nerthern Africa, particularly by French colonists in Algeria. It grows rapidly, so that almost any quantity of the leaves could be obtained. It is said that one man can cut 400 pounds of leaves per day. The extraction of the fibre, which is a very simple process, is usually done by women and children. The fibres are either dried in their natural colour—green— or dyed black to resemble horse-hair, as a substitute for which in upholstery work it is chiefly used. It is exported principally to England, France, Germany, and the United States. The exact date of its introduction is not known. Large quantities of the dried leaves are used for making baskets.

In the early part of 1889 a new fibre from the west coast of Africa was brought to notice at Kew, whence it was received from the Governor of Lagos. It was sent under the name of BOLOBOLO, and is also known in the Yoruba language as AGBONRIN ILASSA. From specimens of the plant which accompanied the fibre, it was proved to be derived from *Honckenya ficifolia*, belonging to the natural order Tiliaceæ. A report on the commercial value of the fibre was furnished by Messrs. Ide and Christie, of Mark Lane, and published in the *Kew Bulletin* for January, 1889. The following is an extract from their report :— " We consider this a very valuable fibre of the jute class,

K

but distinctly superior to the latter in many respects, and more particularly in strength. It is of good length, and well cleaned. If this fibre is capable of being produced in large quantities, there is a very wide field open to it commercially. Its market value would be regulated by that of jute, but in our opinion it would always command a higher price. At to-day's currencies it would sell at £16 per ton in London. We do not think the minimum price would ever fall below £12 ; and if the jute market made a further advance, this Bolobolo fibre might realise £20. If this fibre could be prepared of a whiter colour, it would prove still more acceptable ; but even as it is, we should be very glad to see large quantities placed on this market, where they would sell readily."

Under the name of BOMBAY ALOE FIBRE a sample of a white fibre was received at the Kew Museum at the close of 1888. It was imperfectly prepared, and the price obtained for it in the London market was exceptionally low. The botanical source of this fibre being unknown, application was made by Kew to the India Office to obtain from Bombay plants or material to enable it to be identified, when it proved to be obtained from *Agave vivipara*, a plant closely allied to the common American Aloe. A quantity of about 200 tons was received in the London market in 1889, the prices quoted for which were £12 per ton for good quality and £5 for common. The fibre could only be used for ropes. The subject is fully treated of in the *Kew Bulletin* for March, 1890.

CHAPTER XII.

THE question of the extended cultivation of fodder plants, as well as the consideration of their storage for winter use, has always occupied more or less of the attention of agriculturists, and of late years more practical results have accrued in sending new fodder plants for cultivation into our colonies than in introducing such into our own country.

About fifty or sixty years since, several plants of this character were brought to notice as suitable and very desirable for cultivation in England. The exact dates, however, when they were first proposed it is difficult to fix. About fifty years ago a considerable amount of interest was excited in the GAMA GRASS or BUFFALO GRASS (*Tripsacum dactyloides*) of the Southern States of America. Though it is considered by some a good forage plant, it is somewhat too tender for general cultivation with us.

Aira flabellata, better known as *Dactylis cæspitosa*—THE TUSSOCK GRASS—a strong-growing tufted perennial native of the Falkland Islands, was introduced to Kew in 1842, and seeds were afterwards obtained and freely distributed. It was at one time supposed that it might become thoroughly established in this country, but experience has shown that the climate is too hot and dry. It has, however, succeeded in the West of Scotland, and has become established in the island of Lewis. It is doubtful whether the plant is really adapted for pasture purposes. The tussocks are only formed slowly, so that cattle would soon destroy them, and the plant thus become exterminated.

Under the name of *Bromus Schraderi* a new fodder grass was introduced some fifteen or sixteen years ago. The

K 2

plant, which is now known to botanists as *Ceratochloa unio-loides*, is commonly known as AUSTRALIAN PRAIRIE GRASS. It occurs from Central America to the last Alpine zone of Northern Argentina, and has spread over many parts of the globe. It is described as one of the richest of all grasses, grows continuously, and spreads rapidly from seeds, particularly on fertile and somewhat humid soil. It is a very nutritious fodder and pasture grass, besides which it is said to be very valuable for sowing in coverts, as it entices hares and rabbits into the woods, away from the grain crops.

Prangos pabularia, TIBET HAY.—A perennial, belonging to the Umbelliferæ, forming a stem a few feet high. It is a native of Tibet, as its common name implies, where it is extensively used as a fodder for sheep, goats, and oxen. It was introduced for cultivation as a fodder plant in this country about 1840, but it did not succeed.

Perhaps the most important fodder plant introduced during this century is that which is now so well known as PRICKLY COMFREY. This was first brought to notice in 1877, and advertised as *Symphytum asperrimum*. The history and value of the plant is thus summarised in the Kew Report for 1878 :—" It is apparently identical with a *Symphytum* which has long been naturalised in the neighbourhood of Bath and elsewhere, and which has been identified by botanists with *S. asperrimum*, a native of the Caucasus. Neither the naturalised nor the forage plant appear to be really identical with that species, but the latter has been found by Mr. Baker to agree with *Symphytum peregrinum*, which appears to be not certainly known as wild anywhere, but to be probably a hybrid of garden origin between *Symphytum officinale* and *S. asperrimum*. . . . In England Prickly Comfrey has been found very useful for winter fodder, as it forms large tufts of root leaves, which start into growth early in the year, and bear several cuttings ; it is greedily eaten by animals which refuse ordinary

comfrey, the habit and appearance of which are not very
dissimilar." The acclimatisation of the plant has been

PRICKLY COMFREY (*Symphytum officinale*).

attempted in various parts of the world, including India,
Ceylon, Singapore, Australia, and Queensland, with, however, but little success, as it is more suited for cool or temperate countries.

In 1877 a considerable amount of interest was directed
to the fleshy corollas of the well-known Indian MAHWA
(*Bassia latifolia*). The tree, which belongs to the natural
order Sapotaceæ, is very common in many parts of India,
especially in Bengal, and the flowers are produced in such
large quantities as to cover the ground when they fall; they
are succulent and sweet, somewhat like a raisin in appearance, but with a heavy cloying taste and smell. They are
largely used as an article of food, both fresh and stored for
winter use. In the year previously mentioned (1877) a
quantity of these flowers was sent to England for trial in
feeding cattle, as well as for distilling a spirit from them.
For the first they were reported upon most favourably—the
flesh of pigs fed upon them being said to be especially good—
while for distilling purposes they were said to have yielded
as much as 6·16 gallons of proof spirits per cwt., the flavour
of which was very similar to that of Irish whisky, though
by careful rectification it might be made exceedingly pure and
free from flavour. In India the spirit is manufactured on a
large scale, and it is said that recently the flowers have become a regular article of export from Bombay to France,
where they are distilled, the spirit being put into French
bottles, labelled as French brandy, and exported again to
Bombay. As an article of import to this country, however, Mahwa flowers have not fulfilled what was anticipated
of them.

CHAPTER XIII.

TIMBERS AND HARD WOODS.

THOUGH the extended application of iron during the last twenty years, both for ship and house building, has to some extent supplanted the use of timber, the increased building operations all over the country have caused a continued demand for the various building timbers. The attention of our timber merchants and ornamental wood dealers has not been so much directed to the introduction of new woods as to new sources of supply of existing kinds. The pines and oaks are still the woods mostly in demand for structural purposes, and it is for cabinet-work that most interest is shown in the application of new woods. Notwithstanding all that has been done by the British possessions, as well as by foreign countries, to bring their forest resources prominently forward at the several International Exhibitions since 1851, the result cannot be said to be satisfactory so far as the British timber trade is concerned.

The magnificent collections of Australian timbers that have from time to time been shown, as well as those from the Cape of Good Hope—notably in the Colonial and Indian Exhibition of 1886—have not resulted, as might have been anticipated, in creating a demand for them in this country. It may be thought that a periodical exhibition is not the best means of keeping such things fresh in the minds of those most interested, and to some extent this is true ; but when these collections find a permanent home, always open to the public, as they are at Kew, there can be no such excuse. In the case of Australasian timbers, however, there may be some reason why they have not yet figured as regular articles of import with us, and this is the cost of freight for so long a distance, coupled with the fact that most of the timbers of those far-off colonies are very dense and remarkably heavy. This is, of course, especially the case with the numerous species of *Eucalyptus*, which genus furnishes some of the most characteristic of Australian woods. The hardness of these woods indeed is their special recommendation. One species—namely, the JARRAH (*Eucalyptus marginata*), a native of West Australia—has attracted some attention during the last year or so as a material for paving roadways; and blocks made from it have been laid down by several of the metropolitan vestries, as, for instance, at Islington; King's Road, Chelsea; Westminster Bridge Road; and in the Strand. There are some others that have appeared occasionally in our markets, and ought to be regularly known in the timber trade, if only for cutting into veneers, should the woods be too costly to use in the solid. Of such we may mention MUSKWOOD (*Olearia argophylla*), TASMANIAN MYRTLE (*Fagus Cunninghami*), and HUON PINE (*Dacrydium Franklinii*), all of which have been greatly admired by our ornamental wood dealers; but some system of a demand on this side of the world, and a ready response on the other, seems to be needed to create a trade in these bulky commodities.

So far as woods for cabinet purposes are concerned, though fashion rules the demand in this, as in everything else, there is always a sale for such well-known woods as mahogany (which has been used in this country as a cabinet wood since the middle of the last century), walnut, etc.; and in connection with this it may be worth while here to place on record what has been done in the introduction of the mahogany-tree in India, Ceylon, and Mauritius, so that future generations may draw their supplies of this valuable wood from the East as well as from the West Indies. So far back as 1873 seeds were sent from Kew to India, and in 1879 the cultivation of the tree was referred to as an

GATHERING THE BARK OF CORK TREES.

"accepted success," so that there is apparently no fear of the mahogany supplies failing.

One of the valuable woods that has been introduced to this country within the last fifty years is SABICU, or, as it is sometimes called, SAVICU. It is the produce of *Lysiloma Sabicu,* a leguminous tree of Cuba and San Domingo, from whence it is imported to this country, and latterly in small quantities from the Bahamas. The wood is so hard, dense, and durable, that it was much used at one time in ship-building for keelsons, beams, engine-bearers, stern-posts, etc. It was not much known, however, before 1851, in which year it was used for the stairs of the Great Exhibition; and, notwithstanding the immense traffic upon them, they were found at the close of the exhibition to be but little the worse for wear. In 1879 Bahamas Sabicu wood was first used for weaving shuttles and bobbins, but the demand for this purpose has never been large.

Another building timber of great importance is KAURI (*Dammara australis*). This is a large tree, 100 to 150 feet high; native of the northern Island of New Zealand. Mr. Ransome says, in his report on Colonial Timbers (Colonial and Indian Exhibition, 1886), that "this is undoubtedly the best of all soft woods, being remarkably sound, durable, and straight-grained." It is eminently suitable for doors, straight and circular mouldings, match-boarding, and other joiners' work, as well as for casks and engineers' patterns. The wood has been imported in small quantities for many years, and always meets with a ready sale. It yields a valuable resin known as KAURI GUM. (*See* Resins.)

Probably there is no branch of the subject relating to the supplies of wood or of its utilisation of more importance than that which touches the supply of boxwood, or the discovery of an efficient substitute for engraving purposes. For some years past there has been a gradual falling off in the supplies; indeed, in 1875 it was stated that the boxwood forests of Mingrelia, in the Caucasian range, were almost exhausted, and wood that had been rejected in old forests was being eagerly cut, and purchased at high prices for export to England. The cutting of wood in Abhasia and in all the Government forests in the Caucasus was prohibited, and about the same time a prohibition was issued by the Porte against the cutting of boxwood at Trebizond. The discovery of a wood that might be used as a substitute for box is not a new matter; for many years it has occupied the attention of practical, as well as of scientific men, but up to the present time no wood has been discovered that at all equals box for engraving purposes; so that while other woods may be substituted for the various other uses to which box was at one time largely put—namely, for shuttles, turnery, carving, and ornamental uses—for the best engravings box alone is still in demand. In 1880 some consignments of Indian boxwood were received in the London market; but the difficulty and cost of transit from the Himalayas, where the tree grows, operate against its becoming a regular article of export.

The great increase of illustrated books and newspapers continues to put a heavy pressure on the boxwood resources, so that an efficient, if not a perfect, substitute is as much a necessity as ever.

The following are the names of the principal woods that have been tried and reported upon by practical men during the last few years :—

1. *Acer saccharinum.*—SUGAR or BIRD'S EYE MAPLE. North America. Not favourably reported upon.
2. *Amelanchier canadensis.* — AMERICAN SHADE or SERVICE TREE. Might prove useful.
3. *Brya ebenus.*—COCUS WOOD. Jamaica. Equals bad box.
4. *Bursaria spinosa.*—TASMANIAN BOXWOOD. Found in North, West, and South Australia, Queensland, New South Wales, Victoria, and Tasmania. Equal to common or inferior box.

5. *Carpinus Betulus.*—HORNBEAM.   Britain.   Not very favourably reported upon.

6. *Cornus florida.*—NORTH AMERICAN DOGWOOD.   Rough, suitable only for bold work.

7. *Cratægus oxyacantha.*—HAWTHORN.   Britain.   By far the best wood after box.

8. *Diospyros ebenum.*—EBONY.   Ceylon.   Nearly as good as box in texture ; colour of wood an objection.

9. *Diospyros texana.*—A North American tree.   Nearly equal to best box.

10. *Elæodendron australe.*—Queensland and New South Wales.   Suitable for diagrams, posters, etc.

11. *Euonymus europæus*, var. *Hamiltonianus.*—PAI'CHA.   China, where the wood is much used for carving and engraving.   A useful wood, especially for bold work.

12. *Eugenia procera.*—Jamaica, Antigua, and Martinique.   Suited for bold, solid newspaper work.

13. *Monotoca elliptica.*—New South Wales, Victoria, and Tasmania.   Not very favourably reported upon.

14. *Pittosporum bicolor* and *P. undulatum.*—New South Wales, Victoria, and Tasmania.   Both woods are suitable only for bold outlines.

15. *Pyrus communis.*—COMMON PEAR.   Britain.   Not very well reported upon, but it does well for engraved blocks for calico printers.

16. *Rhododendron californicum* and *R. maximum.*—Both of these have been favourably reported upon from North America.

17. *Tabebuia pentaphylla.*—WEST INDIAN BOX.   West Indies and Brazil.   A fairly good substitute for box.

The most recent substitute for true boxwood that has been brought to notice, and one that at first promised to become of considerable importance, is that known as Cape boxwood. The first notice of this wood was contained in a letter from East London, Cape Colony, in 1885, addressed to the writer, and in the same year about three tons arrived in London. Samples were submitted to several practical men for trial and report, and they all agreed that the wood did not cut smoothly, but was harsh and ragged, and on the whole that it was far inferior to boxwood. The trees were said to be sufficiently abundant in the East London forests to furnish a large supply of wood. Upon receipt of foliage and flowers at Kew, the tree was found to be a new species of *Buxus*, and was named *Buxus Macowani*. The wood has not yet come into general use.

158

CHAPTER XIV.

MISCELLANEOUS PRODUCTS.

UNDER this head are included such products as could not readily be classified under any of the foregoing, but which are—many of them, at least—of great commercial and economic interest. A reference to one trade alone will suffice to prove this—we mean the trade in WALKING-STICKS and UMBRELLA and PARASOL handles; for while at the present time this is one of the great trades of this country, in the early years of the present century it was practically *nil*. There are no published returns showing the importation of raw material used in this trade; but from figures which we have been at some trouble to obtain, it would seem that of rattan canes alone, imported during the year 1886, there were some 1,500 tons, of the estimated value of £30,000, while other canes imported from the East numbered 28,950,000, valued at £94,000; and to these may be added imports from other parts of the world, as Brazil, Algeria, West Indies, France, etc., bringing up the gross total value of rough material to £189,000. Placing this against the value of the imports in 1850 of £1,600, it will be seen what progress has been made in this one trade alone, which deals almost exclusively with produce furnished by the vegetable kingdom. As a further proof of the importance of this trade at the present time, I may mention that Messrs. Henry Howell and Co., of 180, Old Street, City Road, E.C.—the largest firm engaged in this trade in London, and to whom I am indebted for the above facts—constantly employ as many as 530 hands in their establishment. Another trade whose operations are confined almost exclusively amongst plants, and which within the last thirty years has considerably developed as a branch of English commerce is that of

perfumery, for we not only import attar and essential oils
in large and increasing quantities from Roumelia, Singapore,
and other places, but the cultivation of perfume plants in
this country has received more attention; and when we know
that Mitcham lavender and peppermint oils are unequalled
in the markets of the world, there seems no reason why
the cultivation of such plants, and the distillation of their
oils, should not be made specially a home industry.   As an
illustration of the great value of imported perfumery oils,
we will briefly refer to those produced by species of *Andropogon*, which are introductions of the present century.   Thus
LEMON GRASS OIL, the produce of *Andropogon citratus*, was
first imported into London about 1832; while RUSA, or
GINGER GRASS OIL, from *A. Schœnanthus*, was first brought
to notice in 1825, and CITRONELLA OIL, from *A. Nardus*, at a
much more recent period.   Citronella and lemon grass plants
are extensively cultivated in Singapore and Ceylon for commercial purposes, large plantations in the latter place being
devoted to them, and the oil distilled on the spot.   Ginger
grass oil is chiefly distilled in Khandesh, in the Bombay
Presidency.   Twenty-five years ago the export of citronella
oil from Ceylon was 622,000 ounces, of the value of £8,230,
and it has considerably increased since then; besides which
are to be added the still greater exports from Singapore, a
very large proportion of which comes to this country.

As an illustration of what may be done in the utilisation
of waste products, CORK stands forward as a prominent
example.   Fifty years ago the uses of cork—the bark of
Quercus suber—were chiefly as stoppers for bottles, floats
for nets, in the construction of lifeboats, etc.   In 1851, however, the adaptability of cork for very many other domestic
and manufacturing purposes was practically illustrated, and
its uses became wider and more general.   The utilisation of
virgin cork for horticultural purposes does not date back
more than about twenty years; previous to its application for

window-boxes, rockeries, orchid-growing, etc., it was a waste product, as, owing to its irregular growth and porous nature, it is quite useless for stoppers.   Another use, however, has since been found for it—namely, for grinding into powder, and mixing with linseed oil and rubber, in the manufacture of the floor-covering known as linoleum.   In view of the still further extended use of the cork-tree, plants have been intro-duced into India, where they seem to have made healthy and vigorous growth.

VEGETABLE IVORY.—The seeds of *Phytelephas macro-carpa*, a low-growing or almost stemless palm, found on the banks of the river Magdalena, and producing large globular bunches of fruits about the size of a man's head, containing numerous white seeds, which become very hard as they ripen, are extensively used as a substitute for real ivory, chiefly for inlaying, for knobs for drawers, and very largely for coat buttons.   Vegetable ivory is said to have been introduced into Europe about the year 1826, but when it first came into commerce in this country is not accurately known.

During the summer of 1878 London, and indeed the whole of the United Kingdom, was deluged with an enor-mous importation of hats plaited from a kind of sedge. Though they were known to come from China, they soon obtained the name of ZULU HATS, and they found their way even into the remotest villages of the kingdom, being sold at the remarkably low price of one penny each.   So abundant were they indeed that the market became glutted, and the hats were sold for use as strawberry guards in gardens by cutting out the crowns.   The Consul at Ningpo reported that no less than 15,000,000 of these hats, all made by hand, had been exported in one year.   The plant from which they are made, which proved to be *Cyperus tegetiformis*, is cultivated especially for this manufacture in rice grounds, and the hats are made by women and children

The same plant is used for making the Chinese matting which has been imported into this country, and so largely used for bed-room and drawing-room floors during the last six or seven years.

The so-called BRIAR-ROOT PIPES, which have now become such a large article of trade, were first introduced to this country about thirty years ago. For some time their origin was quite unknown, and they were made only in small quantities. A flourishing industry is now established at several places in Italy and France, notably at Leghorn, Siena, and Grossitto. The roots of the "briar," which word is a corruption of Bruyère (*Erica arborea*), are collected on the hills of the Maremma, where the plant grows luxuriantly and attains a great size. When brought to the factory, the roots are cleaned of the earth which is attached to them, and the decayed parts cut away. They are then cut roughly into pipe-shapes, placed in a vat, and gently simmered for twelve hours, by which time they acquire a rich yellowish-brown colour, for which the best pipes are noted. The rough blocks are then put into sacks containing from forty to a hundred each, and sent to France, where they are bored and finished off ready for exportation.

Under the name of LOOFAHS, our chemists have exbited in their shops for the last few years natural flesh brushes, consisting of the vascular tissue of the fruits of *Luffa ægyptiaca*, a climbing cucurbitaceous plant, native of Egypt and Arabia, but grown also in the West Indies and Western Africa, where it is generally known as the Towel Gourd. In the countries where the plant grows, the vascular network of the fruit is commonly used for straining palm wine and other fluids, as well as for scrubbing-brushes, and making light ornamental articles, such as baskets, hats, etc. Quite recently a large factory has been established in Germany for converting the Luffa fruits into useful domestic articles, of which soles or

L

socks to place in boots, to keep the feet dry and warm in winter and cool in summer, are among the most important. They are elastic, and easily washed with soap and water. Saddle undercloths are also made from Luffas, and are intended to supplant the felt cloths hitherto used. They fit the saddle perfectly to the back of the horse, and they prevent the animal remaining wet under the saddle after sweating. Surgical bandage stuffs are also made from Luffas, and are competing with the wood-wool kind introduced some years ago.

The uses to which the Luffas or Loofahs may yet be put are very numerous when we consider that they are obtainable in almost any quantity and at a very low rate ; some bales received in the London market a few years ago having been sold at five fruits a penny.

A new kind of paint or composition, especially intended for coating ships' bottoms to prevent corrosion, was brought to notice, and experiments made with it in Chatham dockyard in 1873, when a sheet of iron coated with the paint was lowered into one of the basins, and after two years' immersion was found to be practically as clean as when first put down. In 1877 a company was formed, under the title of the Protector Fluid Company, for manufacturing this paint on a large scale. The fluid, with which any colour can be mixed, is prepared with the juice of one or more species of *Euphorbia*, collected, it is said, in Natal. The discovery of this property of the *Euphorbia* juice is said to have been made accidentally when cutting plants of Euphorbia in Natal. It was found that the juice adhered firmly, and coated the blades of the knives, thus preserving them from rust. The value of a preservative against corrosion and the attacks of barnacles will be apparent in saving the cost of frequent cleaning, and in maintaining the speed of fast-going vessels.

INDEX.

L 2

PRINTED BY CASSELL & COMPANY. LIMITED, LA BELLE SAUVAGE, LONDON, E.C.

Printed in the United States
By Bookmasters